NUCLEIC ACID HYBRIDIZATION

Editor:

David Rickwood, Department of Biological and Chemical Sciences, University of Essex, Colchester, Essex

CENTRIFUGATION
RADIOISOTOPES
LIGHT MICROSCOPY
ANIMAL CELL CULTURE
GEL ELECTROPHORESIS: PROTEINS
PCR, SECOND EDITION
MICROBIAL CULTURE
ANTIBODY TECHNOLOGY
GENE TECHNOLOGY
LIPID ANALYSIS
GEL ELECTROPHORESIS: NUCLEIC ACIDS
LIGHT SPECTROSCOPY
DNA SEQUENCING
MEMBRANE ANALYSIS
PLANT CELL CULTURE
NUCLEIC ACID HYBRIDIZATION

NUCLEIC ACID HYBRIDIZATION

M.L.M. Anderson

Institute of Biomedical and Life Sciences, Glasgow University, Glasgow, UK

Taylor & Francis
Taylor & Francis Group

LONDON AND NEW YORK

© **Taylor & Francis Publishers Limited, 1999**

First published 1999

All rights reserved. No part of this book may be reproduced or transmitted, in any form or by any means, without permission.

A CIP catalogue record for this book is available from the British Library.

ISBN 1-85996-007-3

Published by Taylor & Francis
2 Park Square, Milton Park, Abingdon, Oxon, OX14 4RN
270 Madison Ave, New York NY 10016

Transferred to Digital Printing 2009

ISBN10: 1-85996-007-3

Production Editor: Andrea Bosher
Typeset by Marksbury Multimedia Ltd, Midsomer Norton, Bath, UK

Publisher's Note
The publisher has gone to great lengths to ensure the quality of this reprint but points out that some imperfections in the original may be apparent.

Contents

Abbreviations

A	adenine
Ab	antibody
AMPPD	disodium 3-(4-methoxyspiro[1,2-dioxetane-3-2'-tricyclo-[3.3.1.1^{3,7}]decan]-4-yl) phenyl phosphate
AP	alkaline phosphatase
ASO	allele-specific oligonucleotide
BCIP	5-bromo-4-chloro-3-indolyl-phosphate
bp	base pair
BSA	bovine serum albumen
C	cytosine
cDNA	complementary DNA
CDP-star	disodium 2-chloro-5-(4-methyoxyspiro[1,2-dioxetane-3,2'-(5'-chloro)tricyclo[3.3.1.1^{7}]decan]-4-yl)-1 phenyl phosphate
CIAP	calf intestinal alkaline phosphatase
cps	counts per second
CSPD	disodium 3-(4-methyoxyspiro[1,2-dioxetane-3,2'-(5'-chloro)tricyclo[3.3.1.1^{3,7}]decan]-4-yl) phenyl phosphate
DABCYL	4-[4'-dimethylaminophenylazo]benzoic acid
dATP	deoxyadenosine triphosphate
dCTP	deoxycytidine triphosphate
ddNTP	dideoxynucleoside triphosphate
dGTP	deoxyguanosine triphosphate
DIG	digoxygenin
DNA	deoxyribonucleic acid
dNTP	deoxynucleoside triphosphate
ECL	enhanced chemiluminescence
EDTA	ethylenediaminetetraacetic acid
G	guanine
HRP	horseradish peroxidase
kb	kilobases

MLP	multi locus probe
mRNA	messenger RNA

NET	1 x NET = 100 mM NaCl, 1 mM EDTA, 10 mM Tris-HCl, pH 7.5, 0.5% SDS
nm	nanometres
nt	nucleotide(s)
NTB	nitroblue tetrazolium

OD	optical density

PCR	polymerase chain reaction

RFLP	restriction fragment length polymorphism
RNA	ribonucleic acid
rRNA	ribosomal RNA
RSP	restriction site polymorphism
RT	reverse transcriptase

SDS	sodium dodecyl sulphate
SLP	single locus probe
SSC	1 x SSC = 0.15 M NaCl, 15 mM trisodium citrate, pH 7.0
SSPE	1 x SSPE = 0.18 M NaCl, 10 mM sodium phosphate, pH 7.4, 1 mM EDTA
SST	1 x SST = 0.15 M NaCl, 15 mM Tris HCl, pH 7.5, 5 mM EDTA

TCA	trichloroacetic acid
TE	Tris EDTA buffer = 10 mM Tris HCl, pH 8.0, 1 mM EDTA
T_m	melting temperature
tRNA	transfer ribonucleic acid
TTP	thymidine triphosphate

U	uracil
UV	ultraviolet

VNTR	variable number of tandem repeats

Preface

The first description of nucleic acid hybridization was by Marmur and Doty in 1961. They established that the two sequences involved in formation of a duplex must have a degree of complementarity and that the stability of the duplex depended on the degree of relatedness. These early experiments provided a means of studying relationships between nucleic acids that was quickly exploited. Application of nucleic acid hybridization to molecular biology has revolutionized our knowledge of gene structure, organization and expression. Many modifications have been made to the original technique that have extended its sensitivity and versatility.

There are many different type of hybridization currently in use. These include solution hybridization, *in situ* hybridization, filter hybridization, PCR and hybridization to DNA chips. Although the range of applications is wide, common features underlie them all.

This book is targetted at the beginner who has little or no knowledge of the principles or practice of hybridization. The field of hybridization is too large to be covered in a single book. Recent companion volumes give comprehensive accounts of PCR and *in situ* hybridization [1–2] and so these topics will not be covered at length here. The book will concentrate on solution and filter hybridization with a final chapter on current developments which includes DNA chips and advances in probe design.

To lay the framework for later discussions, the first chapter describes the structure and properties of DNA, particularly in relation to how they relate to hybridization. Chapter 2 describes the different types of hybridization that are in current use and the type of applications for which they are particularly suited. Some of the drawbacks are also discussed. Chapters 3–5 discuss the principles of reassociation kinetics and how they can be applied to answer biological questions. Many of the concepts that are used in hybridization (e.g. $C_o t$ and $R_o t$, complexity and stringency) originated in studies of solution hybridization and are explained.

Chapter 6 introduces the basic types of filter hybridization on which all the applications are based. There is no single set of conditions that is suitable for all hybridizations. The conditions need to be adjusted according to the purpose of the experiment. Chapters 7–8 discuss the factors which affect hybridization with long and oligonucleotide probes and the choices that have to be made at every step of a hybridization experiment. Chapter 9 discusses choices to be made in selecting and labeling the probe. Chapters 10–13 discuss all the steps in setting up a filter hybridization experiment and give practical details. The methods given are tried and tested so that they should be mastered by a beginner. Chapter 14 describes common problems encountered in filter hybridization, suggests reasons and offers advice on how to deal with them. Chapter 15 discusses a wide range of applications for filter and solution hybridization. The final chapter deals with developments such as hybridization to DNA chips in this fast evolving field.

It is hoped that through understanding the principles of hybridization the beginner can approach scientific literature without feeling intimidated and can begin to read critically. It is intended that the reader should gain sufficient basic knowledge to plan and carry out simple filter hybridization experiments, and should the experiments fail, it is hoped that there will be sufficient information herein with which to diagnose and remedy the problem.

Acknowledgements

I gratefully acknowledge the generosity of the following friends or colleagues who have contributed to this book.

Professor R.L.P. Adams, Dr A.M. Campbell, Professor J.R. Coggins, Miss A. Faichney, Dr G Lanyon, Professor D.M.J. Lilley, Dr A.-M. McNicoll, Dr V. Math and Dr M. Piper.

Figures 15.2 and *15.6* were provided by Dr G. Lanyon and *Figures 6.4* and *13.6* by Dr M. Piper, Boehringer Mannheim Customer Services. *Figures 6.6* and *15.9* are reproduced from Nucleic Acid Research; A practical approach (1985) with permission of IRL Press at Oxford University Press. *Figure 9.1* is reproduced from Gene Probes 1: A Practical Approach (1995) with permission of IRL Press at Oxford University Press.

References

1. **Leitch, A.R., Schwarzacher, T., Jackson, D. and Leitch, I.J.** (eds) (1994) *In Situ Hybridization.* BIOS Scientific Publishers, Oxford.
2. **Newton, C.R. and Graham, A.** (1994) *PCR.* BIOS Scientific Publishers, Oxford.

Safety

Attention to safety aspects is an integral part of all laboratory procedures and both the Health and Safety at Work Act and the Control of Substances Hazardous to Health (COSHH) regulations impose legal requirements on those persons planning or carrying out such procedures.

In this book every effort has been made to ensure that the recipes, formulae and practical procedures are accurate and safe. However, it remains the responsibility of the reader to ensure that all necessary COSHH requirements including completion of forms have been implemented. Any specific instructions relating to items of laboratory equipment must also be followed.

Neither the author, editors nor publishers accept any responsibility for loss or damage occasioned to any person or property through using the materials, instructions, methods or ideas contained herein, or acting or refraining from acting as a result of such use. While the author, editors and publishers believe that the data, recipes, practical procedures and other information as set out in this book are in accord with current recommendations and practice at the time of publication, they accept no legal responsibilty for any errors or omissions, and make no warranty, express or implied, with respect to material contained herein.

1 Introduction to nucleic acid hybridization

Nucleic acid hybridization describes a range of techniques in which single-stranded nucleic acids are incubated under conditions of temperature and ionic strength that favor pairing of complementary bases and duplex formation. Under these conditions a molecule of nucleic acid can reassociate with its complementary sequence even in the presence of a vast excess of unrelated sequences. The techniques are powerful and have had a major impact on our knowledge of gene structure and information flow in the cell. Since they all depend on the base-pairing properties of deoxyribonucleic acid (DNA) and ribonucleic acid (RNA), we will first consider the structure and properties of nucleic acids with special emphasis on properties that affect hybridization.

1.1 Structure of nucleic acids

All nucleic acids are composed of nitrogenous bases, pentose sugars and phosphate. In DNA, the bases are the purines adenine (A) and guanine (G) and the pyrimidines cytosine (C) and thymine (T). In RNA the major bases are adenine, guanine, cytosine and uracil (U). The sugars are deoxyribose in DNA and ribose in RNA (*Figure 1.1*). The basic building brick of a nucleic acid is the nucleotide in which the sugar is attached at carbon atom 1 to the base and at carbon 5 to the phosphate. In nucleic acids, nucleotides are joined into linear polymers via phosphodiester linkages (*Figure 1.2*).

Most DNAs are double-stranded with the individual strands twisting round each other and held together in a helical structure by hydrogen bonding between adenine and thymine and between guanine and cytosine (*Figure 1.3a*). An important consequence of the specificity of base-pairing in DNA is that the sequence of one strand determines the sequence of the other.

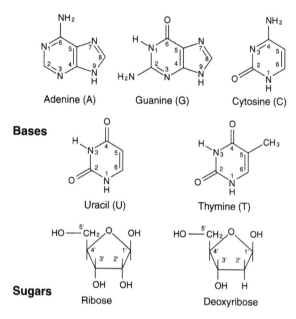

Figure 1.1. Base and sugar components of nucleic acids. Bases are attached via nitrogen 9 of purines and nitrogen 3 of pyrimidines to carbon 1 of the sugar.

Most RNAs are single-stranded, but contain double-stranded regions. the RNA molecule folds back on itself to form hairpin structures in which base-pairing occurs through complementary regions of the same strand. In RNA, hydrogen bonds are formed between A and U and between G and C.

1.1.1 Forces stabilizing DNA structure

Covalent bonds between adjacent nucleotides determine the primary structure of DNA and weak forces determine the three-dimensional shape. These forces are individually weak, but they act co-operatively and are collectively strong enough to maintain a stable structure while being weak enough to allow conformational flexibility. They stabilize the double-stranded helix in several ways.

- *Hydrogen bonds between complementary bases*. The base pairs A:T and G:C have similar dimensions and can fit similarly into the helical space.
- *Hydrophobic effects*. Purine and pyrimidine bases are flat rather hydrophobic molecules. Burying them inside the helix where they are protected from the aqueous environment, increases the stability of the helix.
- *Stacking forces*. When neighboring bases in the same strand are stacked on top of each other, the planar faces of the bases interact

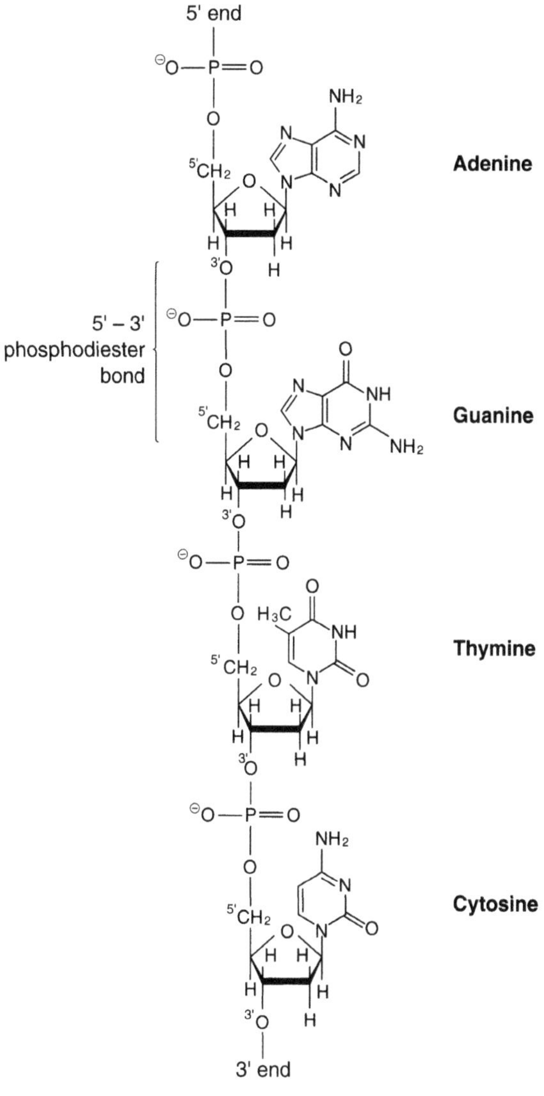

Figure 1.2. Molecular structure of a strand of DNA. Nucleotides are linked by 5′–3′ phosphodiester bonds. A free phosphate group is present at the 5′ end and a free hydroxyl at the 3′ end of the chain.

with each other in a complex way that involves weak van der Waals forces and dipole–dipole interactions.
- *Negative charges.* The negative charges on the phosphate groups of the sugar–phosphate backbone are a potential source of instability in

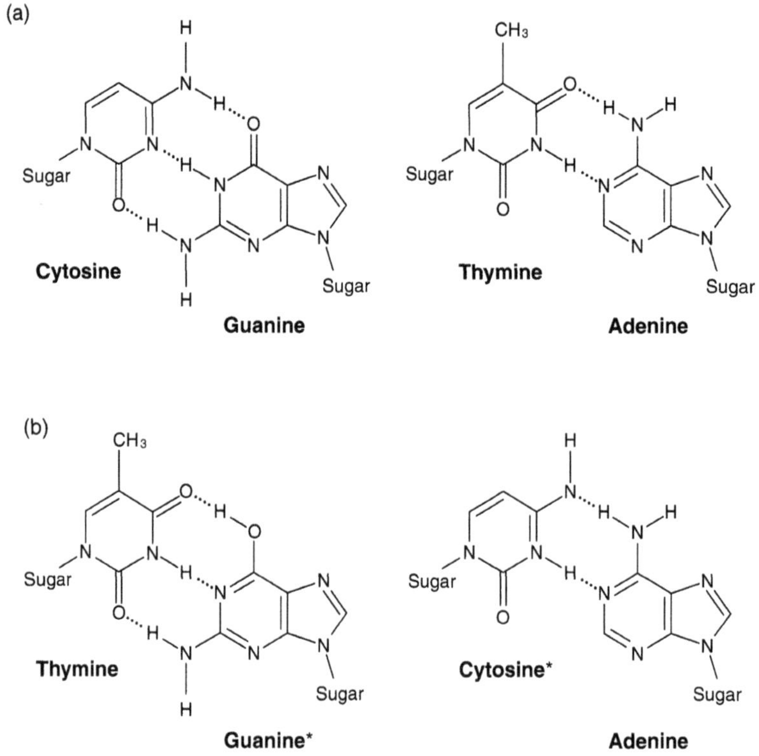

Figure 1.3. Base pairing in DNA. (a) Standard pairs: A pairs with T and C with G. (b) Nonstandard base pairs such as G–T and A–C can be accommodated within the DNA helix, but are less stable than standard base pairs. *Denotes a rare tautomer.

the helix. However, the charge repulsion is overcome by interaction with cations and positively charged side chains of amino acids in proteins.

1.1.2 Mismatched base pairs

The strictness of base-pairing rules (adenine with thymine and guanine with cytosine) produces a regular helix because each base pair is about the same size. Certain mismatches can be tolerated within a double-stranded structure, causing minimal local structural perturbations. Mismatches in base pairing are normally corrected *in vivo*. However, they can easily be created *in vitro* and have a profound effect on the stability of double-stranded nucleic acid. Among nonstandard pairings is guanine with thymine and cytosine with adenine (*Figure 1.3b*).

1.2 Stability of nucleic acids

The noncovalent forces holding double strands together can easily be disrupted. Separation of the strands of double-stranded nucleic acid or the hydrogen-bonded regions of a single-stranded nucleic acid is known as denaturation. Total separation of strands does not occur *in vivo*, but local denaturation can occur in procedures such as DNA replication, transcription and DNA repair. *In vitro*, however, the environment of nucleic acids can be manipulated to alter the stability of nucleic acids and separate the strands.

1.2.1 Effects of pH

Changes in pH can alter the ionization state of bases, sugars and phosphates. At pH values above about pH 11–12 charge repulsion between base pairs causes the complementary strands of DNA to separate. Experimentally this is a very useful means of denaturing DNA. Alkali treatment also causes unfolding of RNA, but must be carried out under very mild conditions as the sugar–phosphate backbone of the RNA chain is hydrolyzed in alkali.

Reducing the pH to below about pH 3 also causes nucleic acid strands to dissociate. However, at low pH the linkage between purines and deoxyribose is broken leading to release of the base. This phenomenon is known as depurination. Pyrimidine/sugar linkages are more stable at low pH than those involving purines. Because of the loss of purines at low pH, acid treatment is not used experimentally for denaturation of nucleic acids.

1.2.2 Effects of temperature

One of the simplest ways of denaturing nucleic acids is to raise the temperature until all the hydrogen bonds and stacking forces have been broken. The strands then dissociate or 'melt'. Single-stranded nucleic acid absorbs UV light more strongly than double-stranded DNA, a phenomenon called hyperchromicity, so as regions of the DNA melt, the absorption of light by those regions increases. The increase in absorption is proportional to the extent of denaturation so dissociation can be followed by monitoring absorbance as the temperature is being raised (*Figure 1.4a*).

When a solution of DNA is heated, there is initially little change in absorbance as the temperature increases, until over a very short temperature range there is an abrupt increase of about 40% followed by no further change. These three phases correspond to native double-

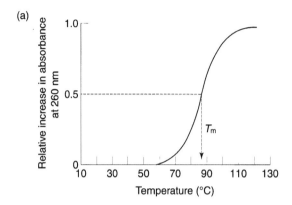

Figure 1.4. (a) Thermal melting curve for DNA. As the DNA is heated, the weak forces holding the double-stranded DNA together are broken. As regions of the DNA melt, the absorption of light of these regions increases so that the increase in absorption is proportional to the extent of denaturation. At T_m, the change in absorption is half the total change.

stranded DNA, breaking of the weak forces holding the two strands together and single-stranded DNA. If the DNA solution is heated very slowly, the structural changes are co-operative and reversible. The midpoint of the increase in absorption is defined as the melting temperature (T_m).

The T_m of a nucleic acid is a measure of its stability – the more stable a molecule is, the higher is the temperature required to dissociate the strands. T_m depends markedly on base composition. A GC-rich DNA melts at a much higher temperature than an AT-rich one because there are three hydrogen bonds to be broken for every G:C complementary base pair whereas there are only two for every A:T base pair. In addition, the stacking forces between GC neighbors is greater than those between AT neighbors.

So, more energy in the form of heat must be supplied to melt the strands of a GC-rich DNA. The dependency of T_m with GC content is linear and increases by about 0.4°C for each percent increase in GC content (*Figure 1.4b*). Under conditions where mammalian DNA (GC content = 40%) melts at about 87°C, a DNA with GC content of 50% will melt at 91°C. For an unknown sample, determination of the T_m can be used to determine the (G+C) content.

Other factors such as the ionic strength of the solution in which the DNA is dissolved affects the T_m (*Figure 1.4b*). At low ionic strength, the negative charges on phosphate groups repel each other so the duplex is

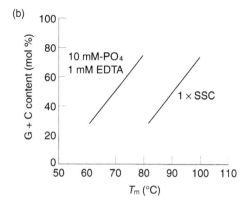

Figure 1.4. (b) Dependence of melting temperature T_m on guanine + cytosine (G + C) content of DNA. There is a linear relationship between the G:C content of DNA and the T_m. Salt stabilizes the double-helix so as the ionic strength is lowered, the T_m decreases.

less stable than in high ionic strength solutions where the salt shields the electrostatic forces and stabilizes the helix. Less energy is, therefore, required for strand separation and the DNA melts at a lower temperature in low ionic strength solutions.

At the T_m of a long nucleic acid, 50% of the base pairs have been broken and the molecules contain both single-stranded and double-stranded regions (*Figure 1.5a*).

There are important differences between the T_m of a long duplex and that of a hybrid containing an oligonucleotide. In the latter the T_m is much lower than that for long DNA because there are far fewer weak forces to be disrupted. Oligonucleotide hybrids are too short to contain both single and double-stranded regions. Their T_m is the temperature at which 50% of the molecules have dissociated into single-strands (*Figure 1.5b*). At the T_m there is equilibrium between single-strands and duplexes.

RNA:RNA and RNA:DNA duplexes can be formed *in vitro*. By virtue of the angle at which the bases are tilted to the long axis of the RNA, these duplexes are more compact than DNA:DNA duplexes. As a consequence the RNA-containing molecules are more stable and have higher melting temperatures than DNA:DNA duplexes with the same base sequence. In general, the order of stability is (from greatest to least stable): RNA:RNA > RNA:DNA > DNA:DNA.

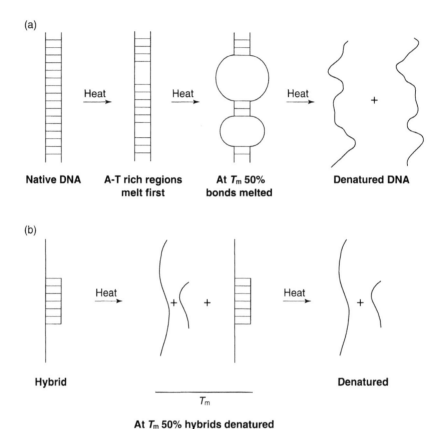

Figure 1.5. (a) Heat denaturation of 'long' DNA. As the temperature is raised, AT-rich regions of DNA melt first. At the T_m half the forces stabilizing the helix have been broken. (b) Heat denaturation of hybrids containing oligonucleotides. Oligonucleotide-containing hybrids melt over a very sharp temperature range. At T_m half the hybrids have dissociated.

Prolonged incubation of nucleic acids at high temperatures causes the sugar–phosphate backbone to break. RNA in particular undergoes rapid nonenzymic hydrolysis at high temperature, being degraded within 2 h at 80°C.

Substances such as urea, dimethyl sulfoxide (DMSO) and formamide break hydrogen bonds and thus reduce the T_m of nucleic acids. In their presence the T_m is lowered by an amount that depends on the concentration of denaturant. This property is very useful experimentally because when denaturants are present, the temperatures required for strand dissociation are lower and nucleic acids do not suffer the thermal degradation that occurs at higher temperatures.

1.3 Renaturation and hybridization

When heat-denatured DNA is cooled slowly, the complementary strands reassociate to form a double-stranded molecule. This is a two-step process (*Figure 1.6*). First, the DNA strands collide randomly with each other and during the process may form imperfectly base-paired stretches. However, these will not be very stable and will quickly dissociate. More collisions will occur until eventually one successfully brings the bases of complementary strands into correct register. This event is known as nucleation. The second step is a rapid zippering up that produces a correctly base-paired duplex. In most reassociations and hybridizations the rate-limiting event is nucleation and such processes are said to be nucleation limited. Slow cooling or incubation at a carefully determined temperature allows many initial base pairings to be sampled and broken until the most stable pairing is retained. On complete renaturation, the properties of the original DNA molecule are restored.

If denatured DNA is cooled quickly (e.g. by plunging the DNA-containing tube in ice), the single strands do not reassociate because there is not enough time for the correct nucleation event to occur.

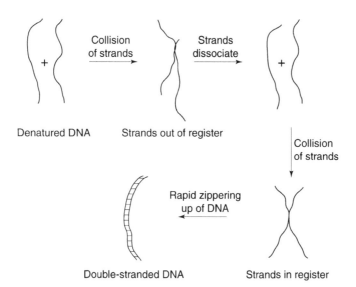

Figure 1.6. Reassociation of DNA. Single strands of DNA undergo numerous random collisions until eventually a collision occurs that places complementary sequences in the correct register. Formation of a base-paired duplex quickly follows.

Instead the strands are trapped either in single-stranded or in imperfectly base-paired structures.

If denatured DNA or RNA is incubated so that the original base pairs are reformed, the nucleic acid is said to have renatured or reassociated. If the complementary strands come from different sources such as DNA from different species or a mixture of DNA and RNA, the process is known as hybridization.

Complementary strands do not have to be perfectly matched to form duplexes. Two strands of DNA with very similar, but not identical sequences will form hybrid double-stranded molecules, but these hybrids will be less stable (will have a lower T_m) than the perfectly matched parental DNAs. This is because in the mismatched hybrids there are fewer hydrogen bonds to be broken in order to dissociate the strands. At the sites of mismatching some combinations of bases are more stable than others, for example, G in one strand opposite T in the other is more stable than A opposite C.

The *in vitro* formation of double-stranded molecules from complementary strands underlies all techniques based on nucleic acid hybridization. The following chapter introduces the main types of hybridization that are currently in use.

2 Types of hybridization and uses of each method

The main types of hybridization used today are liquid hybridization, filter hybridization, the polymerase chain reaction (a specialized form of solution hybridization), *in situ* hybridization and hybridization to 'chips' (a specialized form of hybridization on a solid surface).

2.1 Solution hybridization

In solution hybridization, the reacting species (denatured, single strands of nucleic acid) are free in solution and are incubated under conditions that favor hybrid formation. To detect and monitor hybrid formation, advantage is taken of the difference in physical properties between single- and double-stranded nucleic acids. Duplex formation is usually measured by hypochromicity (reduction in the absorbance at 260nm as double-stranded regions are formed) or selective binding of single and double strands to hydroxyapatite columns.

Sequences present at higher concentration reassociate faster than those present at lower concentration. This is the basis of reassociation kinetics (the rate at which denatured DNA becomes double-stranded) which is used to determine if DNA samples contain sequences that are repeated relative to others. This approach allows the number of copies of repeated sequences, their size and how they intersperse with unique sequences to be determined. The size of the genome can also be deduced. Repeated and single copy sequences can be isolated. The degree of relatedness between sequences can be analyzed by measuring the T_m of reassociated DNA and this approach can be extended to study the relatedness of sequences between genes in different species.

The number of different species of RNA in a cell and the number of copies of each species can be determined. The proportion of the DNA that is transcribed into mRNA can be derived from saturation hybridization studies. The degree of overlap of RNAs expressed in different cell types can be measured. Solution hybridization is also used extensively for mapping the termini of transcripts in studies of gene organization.

Solution hybridization is the standard method for studies which determine the numbers of copies of sequences and their relatedness. However, the procedures are tedious and not well suited to carrying out many different analyses at once. Furthermore, problems arise when denatured DNA from two different sources are incubated together. Providing that the two DNAs have significant sequence homology they will form hybrids. However, there will be a competing reaction from the two parental DNAs which will reassociate to reform the original molecules. Competing reactions occurring at the same time makes it very difficult to interpret results. The problem can be overcome by immobilizing one of the reacting species and this is the basis of filter hybridization

2.2 Filter hybridization

In filter hybridization, single-stranded DNA or RNA is bound to an inert surface and is hybridized to a nucleic acid – the probe – added in solution. The probe carries a reporter molecule or label which is the basis for the subsequent detection of hybrids. After the filter-bound sequences have been incubated with the probe, nonreacting probe is removed by washing and the hybrids which remain on the filter are detected by means of the reporter molecule.

The nucleic acid bound to the filter can come from various sources, for example plates containing recombinant bacteriophage or plasmids, DNA or RNA size fractionated on gels, solutions of crude or purified nucleic acids applied in small dots.

Filter hybridization is widely used to isolate phage and plasmids containing sequences of interest from recombinant libraries. Through analysis of size-fractionated DNA and RNA the size and number of hybridizing species in samples can be determined. This approach is very useful in mapping of DNA. Filter hybridization is capable of great discrimination and can detect single base changes in nucleic acid. This application is widely used in medical research, to detect mutations that cause disease.

Filter hybridization is very versatile and through dot blot hybridization is ideally suited to analyzing many samples at once. It can be used quantitatively in conjunction with a calibration curve for estimating copy number of sequences, but is more commonly used semiquantitatively to compare the relative amounts of sequences in different samples. The main disadvantage of filter hybridization is that it is much slower than solution hybridization and the times required for hybrid formation to go to completion are often impossible to attain. Thus, it is a less useful technique than solution hybridization for analyzing rare sequences.

2.3 *In situ* hybridization

In situ hybridization is a powerful method used to locate nucleic acid sequences in histological and cytological preparations of tissues, organelles, cells and chromosomes [1]. Before hybridization samples are pretreated in such a way that, ideally:

- The morphological features of the tissue/cell/chromosome are retained;
- The nucleic acid is neither extracted nor modified;
- The localization of the nucleic acid is unchanged;
- The probes and detection agents gain access to the nucleic acid.

These aims are very difficult to attain because they can be conflicting. An excellent histological preparation may not allow penetration of the probe and a preparation that allows easy access of reagents, may have lost morphological features. It is, therefore, a major challenge to find an acceptable compromise [2].

Applications include: identifying the location of genes on normal and aberrant chromosomes; identifying the sites of gene expression; determining levels of transcription and how these change with development; detecting viral and pathogen infection. *In situ* hybridization can be used to detect two or more target nucleic acids simultaneously and can detect both a particular transcript and the protein encoded by it. The method is particularly suited to detecting a sequence that is present in only a few cells within a large population, for example chromosomal translocations in residual disease.

For many applications such as detecting genes on chromosomes, the use of sophisticated microscopes is required and may be beyond the resources of many laboratories.

2.4 Polymerase chain reaction

The polymerase chain reaction (PCR) is a procedure for generating large amounts of a specific DNA target in an enzymatic reaction [3]. The procedure involves a cycle of steps in which the starting DNA is denatured, oligonucleotide primers bordering the region to be amplified are annealed and the primers are extended by a thermostable DNA polymerase (*Figure 2.1*). The cycle is repeated some 20–40 times leading

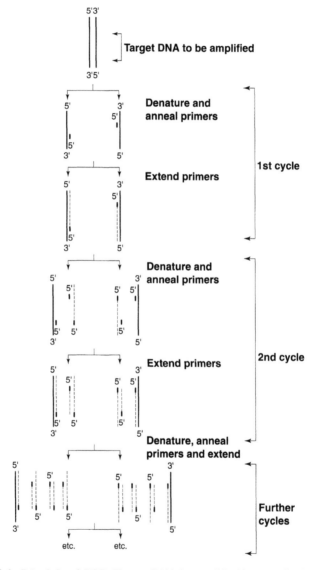

Figure 2.1. Principle of PCR. Target DNA is amplified in a cyclical process that involves denaturation of DNA, cooling to allow annealing of specific primers and extension of primers using a thermostable DNA polymerase and dNTPs. After each cycle the amount of target DNA doubles. The products of the first two complete rounds of PCR are shown.

to amplification of the target sequence by up to 10^5-fold. The amplification is complete within a few hours which compares with days, weeks or possibly even months for isolating the same sequence by cloning.

The ability of PCR to amplify single sequences from a background of many others means that sequences in the original sample which were too rare to be detected by other methods become major species after amplification. This allows problems to be tackled which were intractable a few years ago because the sequences of interest were too scarce. By modifying the primers, PCR can be used to facilitate cloning the products of amplification, generate mutations, detect mutations, join two sequences without the need for restriction enzymes, sequence DNA and compare the DNA in different individuals.

The impact of PCR has been enormous. It is not only used as a tool in basic research, but also for applications ranging from prenatal diagnosis and analysis of carrier status in genetic diseases to forensic analyses, pathogen detection and archeological studies. The techniques are constantly being refined and the range of applications extended. Current developments include modifying the procedure to allow a greater degree of automation.

The strength of PCR in amplifying small quantities of DNA is also one of its weaknesses. Great care must be taken to prevent introduction of any extraneous nucleic acid because it may be amplified. Each step in PCR needs to be carefully optimized or the wrong sequence may be amplified. If this happens early in the procedure, nonspecific sequences may become a significant proportion of the product.

For further details see references [4,5].

2.5 DNA chips

DNA chips consist of large numbers of oligonucleotides or cloned DNA sequences attached to a small surface in such a way that that each sequence has a known position. Many thousands of sequences can be immobilized on a surface smaller than a microscope slide. Hybridization usually takes place using probes labeled with fluorescent dyes. After hybridization, the position of hybrids is detected automatically by a computer-controlled reader. Applications include sequencing and detecting mutations.

The procedure is well-suited to automation, but the equipment required is specialized and is expensive. So this technique may currently be beyond the resources of many laboratories.

2.6 Use of more than one hybridization technique

Experiments can involve more than one method of hybridization. For example:

- *In situ* hybridization may involve PCR on a microscope slide in order to amplify the hybridization signal.
- The products of PCR may be a mixture of sequences of the same size that have to be resolved by cloning and this may involve identifying the desired sequence by filter hybridization.
- Repetitive human DNA isolated by solution reassociation can be used as a molecular probe in filter hybridization, e.g. to identify clones containing human DNA from rodent–human interspecies hybrids.

References

1. **Leitch, A.R., Schwarzacher, T., Jackson, D. and Leitch, I.J.** (eds) (1994) *In Situ Hybridization*. BIOS Scientific Publishers, Oxford.
2. **McNicoll, A.M. and Farquharson, M.A.** (1997) *J. Pathol.* **182:** 250–261.
3. **Saiki, R.K., Scharf, S., Faloona, F., Mullis, K.B., Horn, G.T., Elrich, H.A. and Arnheim, N.** (1985) *Science* **230:** 1350–1354.
4. **Newton, C.R. and Graham, A.** (1994) *PCR*. BIOS Scientific Publishers, Oxford.
5. **McPherson, M.J., Hames, B.D. and Taylor, G.R.** (eds) (1995) *PCR 2: A Practical Approach*. IRL Press, Oxford.

3 Solution hybridization: reassociation of DNA

3.1 Reassociation kinetics

When two complementary strands of DNA are incubated under appropriate conditions, they will reassociate to form a duplex. The rate of reaction as measured by the disappearance of single strands (into double-stranded DNA), is described by the equation:

$$\frac{-dC}{dt} = k \ C \times C = k \ C^2 \tag{3.1}$$

where C is the concentration in nucleotides of the single strand (mol l^{-1}), t is the time (seconds) and k is the rate constant for a second order reaction $(l \ s^{-1} \ mol^{-1})$.

Equation 3.1 shows that the rate of reassociation is dependent on the following:

- **The concentration of DNA.** The higher the concentration of DNA, the faster the reaction proceeds. This is because at higher concentrations of DNA, there are more copies of the complementary strands than at lower concentrations and therefore collisions and nucleation events occur more frequently.
- **The reaction conditions.** The rate constant, k, is affected by variables such as temperature, salt concentration and DNA fragment size. To allow results from different laboratories to be compared, rates of reaction are usually expressed under standard conditions. Correction factors are applied if necessary (see Chapter 5).

It can be shown (Appendix B) that the concentration of DNA remaining single-stranded, C, at time t, is related to the total concentration of DNA, C_o, by the equation:

$$\frac{C}{C_o} = \frac{1}{1 + kC_o t} \tag{3.2}$$

When the reaction is half complete, at time, $t_{1/2}$,

$$\frac{C}{C_o} = \frac{1}{2}$$

and

$$C_o t_{1/2} = \frac{1}{k} \qquad (3.3)$$

Thus, $C_o t_{1/2}$ is inversely proportional to the rate constant and is a measure of the rate of reaction.

3.2 Experimental analysis of a reassocation reaction

The value of $C_o t_{1/2}$ can be determined experimentally by measuring the extent of reaction at different times of reaction. The following steps are generally involved.

1. DNA is broken into small fragments of several hundred nucleotides in length by procedures such as shearing, sonication, high speed blending or digestion with a restriction endonuclease that cuts frequently.
2. Fragments are denatured by heating briefly at about 90°C.
3. Dissociated fragments are incubated under conditions that facilitate reassociation (see Chapter 5).
4. The extent of reassociation (i.e. the fraction of the input single-stranded DNA that has become double-stranded) is measured at intervals.
5. The extent of reassociation is plotted against the time of reassociation or more commonly against $C_o t$ which is the product of the total concentration of DNA and the time.

There are three main methods for following the progress of reassociation.

- *Optical methods*. As noted in Section 1.2.2, nucleic acid denaturation is accompanied by an increase in UV absorption (hyperchromicity). As reassociation takes place, there is a corresponding decrease in the UV absorption (hypochromicity). So DNA reassociation can be followed by measuring the change in absorption at 260 nm with time.

  Measurements are carried out in a temperature-controlled cuvette in a spectrophotometer which is capable of providing continuous readout of UV absorption. The temperature is first raised to denature the DNA and to provide a maximum optical density reading against which all the other readings are compared. The temperature is then reduced rapidly to the incubation temperature and maintained there for the remainder of the experiment. The change in UV absorption with time is recorded.

- **Hydroxyapatite chromatography.** Double-stranded nucleic acid binds firmly to hydroxyapatite (a form of calcium phosphate) at 50°C in 0.14 M sodium phosphate buffer or 60°C in 0.12 M sodium phosphate buffer whereas single strands do not. So, to follow reassociation, aliquots of the reassociation mixture are removed at intervals and applied to a hydroxyapatite column which is maintained at 50° or 60°C. The single-stranded nucleic acid does not bind and is collected in the flow through. Double-stranded DNA is then eluted from the column by raising the salt concentration to 0.4 M sodium phosphate.

If large amounts of nucleic acid are used, the eluates can be quantitated by measuring the absorbance at 260 nm. If the amounts are small, one of the reactants usually contains a radioactive label and is quantitated by liquid scintillation counting. The proportions of starting nucleic acid that are single- or double-stranded are compared at different times of reaction to allow the rate of formation of hybrid to be determined.

- **Nuclease S1 digestion.** Under carefully controlled conditions, nuclease S1 digests single-stranded regions or single-stranded tails of nucleic acid and leaves duplexes intact. The nuclease-resistant nucleic acid remaining is a measure of duplex formation. Nuclease S1 treatment can therefore be used to monitor the progress of a hybridization reaction.

At different times of reaction, samples of the hybridization mixture are removed and digested with nuclease S1 so that only perfect double strands remain. The input nucleic acids are usually radioactively labeled so that after digestion the product can be quantitated by liquid scintillation counting.

It should be noted that these three methods do not actually give the same information. Hydroxyapatite analysis measures the proportion of nucleic acid that is fully single-stranded (DNA that is fully double-stranded or part double-stranded and part single-stranded binds to the column). Nuclease S1 digestion measures the fraction of nucleotides that are completely double-stranded and should give results similar to those obtained from optical measurements.

Figure 3.1 shows the results of a simple reassociation experiment. Two different concentrations of a simple prokaryotic DNA have been denatured and incubated under conditions that favor duplex formation. The extent of reaction was measured by hypochromicity throughout the incubation. The same data are plotted in different ways in *Figure 3.1(a), (b)* and *(c)*.

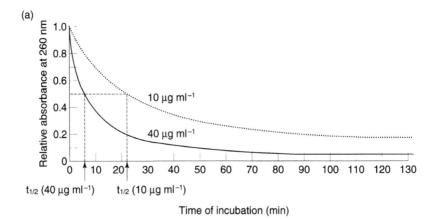

Figure 3.1. Dependence of reassociation time on the concentration of DNA. Two different concentrations of bacteriophage T7 DNA are fragmented, heated to 95°C to separate strands then incubated at 60°C to allow reassociation to occur. The progress of reassociation is followed by measuring the change in absorbance at 260 nm with time. The reaction is complete when the relative absorbance has fallen to 0.73. (a) Change in A_{260} is plotted as a function of the time of reassociation. The time required for the reaction to be half complete ($t_{1/2}$) occurs when the absorption change is half maximal.

In *Figure 3.1(a)* the extent of reassociation is plotted against the time of incubation. The more concentrated DNA ($40\,\mu g\,ml^{-1}$) reassociates faster than the more dilute ($10\,\mu g\,ml^{-1}$) in agreement with the predictions of Equation 3.1. Experimentally, it is impossible to measure the very instant at which the reaction is complete, i.e. when all DNA has become double-stranded, but the time when 50% is in duplex form can easily be determined from the graph. The DNA that is at four times the concentration of the other, reaches 50% completion of reassociation four times quicker than the more dilute solution.

A more common and more informative way of expressing the same data is shown in *Figure 3.1(b)*. Here the extent of reaction is plotted as a function of $C_o t$ instead of time alone and the scale on the abscissa is logarithmic instead of linear. The data from both concentrations of DNA fall on a single curve and there is a single $C_o t_{1/2}$ value, that is the value of $C_o t$ at which the reaction is half complete. In fact it does not matter what concentration of a given DNA is reassociated, its $C_o t_{1/2}$ value will remain constant provided that exactly the same reaction conditions are used (i.e. same temperature, salt concentration, fragment size, etc.). This property is very useful experimentally because it means that by using more than one starting concentration of DNA, the range of $C_o t$ values

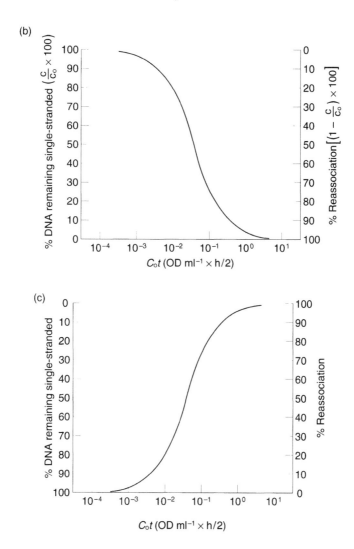

Figure 3.1. (b) Change in A_{260} is plotted as a function of the product of the total concentration of DNA and time of incubation i.e. ($C_o \times t$). (c) The same data as in (b) is plotted, but the scale on the ordinate is in the opposite orientation.

over which reassociation can be sampled is extended. This practice is extensively used when eukaryotic genomic DNA is reassociated.

Note that the curve in *Figure 3.1(b)* is sigmoidal and symmetrical in shape. Between 10% and 90% reassociation takes place over a 100-fold range of $C_o t$ values. This behavior is characteristic of a DNA in which all the fragments are present at the same relative frequency (i.e. the DNA lacks repeated sequences).

In practice, reassociation kinetics are usually followed by preparing a curve as in *Figure 3.1(b)*. This is called a C_ot or Cot curve. Several different starting concentrations of denatured DNA are reassociated under identical conditions and the proportion of DNA that remains single stranded is measured throughout the incubation.

3.2.1 Presentation of data

It should be noted that in *Figure 3.1(b)* the curve has been plotted with two different scales on the ordinate. The left-hand ordinate is labeled '% DNA remaining single-stranded' and 0 is at the origin. The ordinate on the right-hand side is labeled '% hybridization' (which is equivalent to [100% − % DNA remaining single-stranded]) and 100% is at the origin. The curves are identical.

However, in the scientific literature, it is quite common to find C_ot curves plotted with the scales written in the opposite direction. This gives the curve a completely different appearance (*Figure 3.1(c)*) which can be quite disconcerting to the beginner.

In scientific publications, $C_ot_{1/2}$ is expressed in terms of moles of nucleotides per litre × seconds. This is unwieldy and for everyday use a very similar numerical value can be obtained from the expression [1]:

$C_ot = A_{260}$ of DNA per ml × h /2.

This is based on a molar extinction coefficient of 7200 for alkali-denatured DNA.

Some very useful deductions can be made from Equations 3.2 and 3.3.

- If $C_ot_{1/2}$ or the C_ot value at any C/C_o is known, the relative times for different degrees of completion of hybridization can be calculated from Equation 3.2.

For			
	25% hybridization,	$C/C_o = 0.75$	$kC_ot = 0.33$
	50%	$C/C_o = 0.5$	$kC_ot = 1$
	75%	$C/C_o = 0.25$	$kC_ot = 3$
	90%	$C/C_o = 0.1$	$kC_ot = 9$
	95%	$C/C_o = 0.05$	$kC_ot = 19$
	99%	$C/C_o = 0.01$	$kC_ot = 99$

Since kC_o is constant within a set of hybridizations, kC_ot is a measure of the time of incubation required. These calculations show that the relative times of incubation increase markedly for higher degrees of completion of hybridization. For a reaction to proceed to 90% completion, incubation must be carried out for nine times as long as the time required for the reaction to proceed half way.

- Since $C_0t_{1/2}$ is the product of the initial concentration of DNA and the time for the reaction to proceed half way, it follows that the higher the $C_0t_{1/2}$, the slower the reaction.
- If the rate constant k is known, then the value of $C_0t_{1/2}$ can be derived, and *vice versa* (see Equations 3.4 and 3.5).

3.3 Relationship between the rate of reassociation and the sequence complexity of DNA

The term complexity or sequence complexity refers to the total length of different sequences present in a sample of DNA or RNA. In an organism in which there is no sequence repetition, the complexity of DNA is the same as the genome size. Prokaryotes have no significant sequence repetition so the complexity is the same as the genome size.

If an organism contains some sequences that are present only once in the haploid genome and others that are present in more than one copy, both types of sequence contribute to the complexity. The complexity of a family of repeated sequences is the same as their canonical sequence, but the number of times a sequence is repeated is irrelevant, for example if a sequence of 300 bases is repeated 1×10^6 times, the complexity is 300. Suppose a sample contains single-copy DNA with a total length of 1000 nucleotides (nt) and repeated sequences of canonical length 300 nt, the complexity of the DNA is 1000 + 300 = 1300 nt.

The complexity of mRNA is far lower than that of genomic DNA because only a fraction of the genome is transcribed. Similarly, the complexity of a recombinant cDNA library is far lower than that of a genomic library.

Kinetic complexity is the complexity as determined by reaction kinetics. $C_0t_{1/2}$ is proportional to the complexity of DNA:

Consider equimolar solutions of bacteriophage T4 DNA and *Escherichia coli* DNA. The genome size of T4 is 1.7×10^5 bp whereas that of *E. coli* is 4.2×10^6 bp. (The genome sizes here are the same as the complexities since both are simple DNAs and lack repetitive sequences.) The solutions will contain the same total number of nucleotides, but there will be 25 times more copies of T4 DNA in the first solution than there will be of *E. coli* DNA in the second. So, in reassociation reactions, the T4 DNA will reassociate 25 times faster than the *E. coli* DNA and its $C_0t_{1/2}$ will be 1/25 of that of *E. coli* DNA. If they were required to reassociate at the same rate, the concentration of *E. coli* DNA would have to be increased 25-fold, but its $C_0t_{1/2}$ value would not change (Section 3.2).

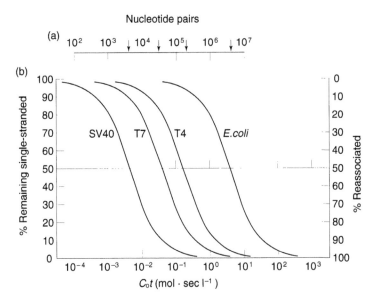

Figure 3.2. C_ot curves of several simple DNAs. (a) A scale relating the molecular size of DNA to $C_ot_{1/2}$. (b) The fraction of DNA remaining single-stranded is plotted against C_ot. The rate of reassociation is inversely proportional to the size of the reassociating DNA.

The effect of complexity on the reassociation rate can be shown experimentally by preparing C_ot curves for different DNAs reassociating under identical conditions (*Figure 3.2*). For prokaryotic DNAs, simple C_ot curves are obtained and their shape is very similar to that in *Figure 3.1(b)*. The $C_ot_{1/2}$ values for different DNAs differ markedly such that those DNAs with higher $C_ot_{1/2}$ values have higher genome size. By analyzing pure DNA of unique sequence and known size, a scale relating $C_ot_{1/2}$ to molecular size can be obtained (*Figure 3.2(b)*).

Two important relationships between complexity and reassociation rate emerge:

- the rate of reassociation is inversely proportional to the complexity of the reassociating DNA;
- all DNAs of the same complexity share the same $C_ot_{1/2}$.

Since the reassociation of DNA of any genome displays a $C_ot_{1/2}$ that is inversely proportional to its complexity, the complexity of any genome can be determined by comparing its $C_ot_{1/2}$ with that of a standard DNA of known complexity. Note, however, that this is only valid if the reassociation of the standard and unknown DNA take place under identical conditions, because reaction conditions affect the rate of reassociation (Chapter 5).

The genome of *E. coli* is often taken as a standard.

$$\frac{C_o t_{1/2} \text{ (DNA of any genome)}}{C_o t_{1/2} \text{ (E. coli DNA)}} = \frac{\text{Complexity of any genome}}{4.2 \times 10^6 \text{ bp}}$$

Under conditions of $1\,\text{M}$ [Na$^+$] and an incubation temperature of $25°\text{C}$ below the T_m, the rate constant, k, is related to the complexity, N, by the empirical relationship:

$$k = 3.5 \times 10^5 \times L^{0.5} \times N^{-1} \tag{3.4}$$

where L is the average length of fragments (nt), and N is the complexity (nt) [2].

By combining Equations 3.3 and 3.4, the relationship between $C_o t_{1/2}$ and complexity is derived:

$$C_o t_{1/2} = \frac{N}{3.5 \times 10^5 \times L^{0.5}} \tag{3.5}$$

3.4 $C_o t$ curves of eukaryotic DNA show the presence of repeated sequences

---

The haploid gene content of eukaryotes is much larger than that of prokaryotes. For example, the haploid genome of mammalian cells (3×10^9 bp) is about 710 times larger than the *E. coli* chromosome (4.2×10^6 bp). If all the DNA in mammalian genomes were single-copy (i.e. there were no repeated sequences), the $C_o t_{1/2}$ for reassociation of DNA would be 710 times that for *E. coli* DNA or 2840 mol s l^{-1} under the conditions in *Figure 3.2(b)*. For a concentration of mammalian DNA of $50\,\mu\text{g ml}^{-1}$, it would take about 8 months for the reassociation to reach half completion and over 13 years to be complete!

In the early days of reassociation kinetics, such reasoning led to the expectation that denatured, eukaryotic DNA would fail to reassociate in a meaningful time. It was a surprise to find that not only could reassociation be detected, but some of the DNA reassociated much faster than *E. coli* DNA [3]. A hypothetical and idealized eukaryotic $C_o t$ curve is shown in *Figure 3.3*. Reassociation takes place over a very large range of $C_o t$ values – up to nine orders of magnitude. This is much larger than the thousandfold range of $C_o t$ values seen for reassociation of prokaryotic DNA.

The key to interpreting eukaryotic $C_o t$ curves is that sequences present at high concentration reassociate faster than those present at low

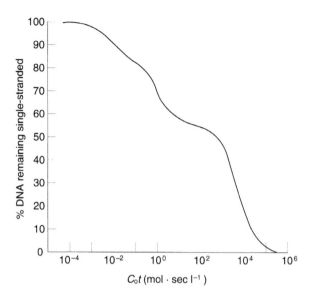

Figure 3.3. C_ot curve of DNA from a higher eukaryote. The curve can be resolved into three components that differ in the rate at which they renature. The faster a component renatures, the more copies of it are present.

concentration. So the sequences which reassociate most rapidly must be present in very high copy number. The overall shape of the C_ot curve represents the sum of a number of independent reassociations each of which has its own characteristic $C_ot_{1/2}$ value.

The curve in *Figure 3.3* can be resolved into three components. The first component to reassociate, the fast component, renatures between C_ot values of 10^{-4} and 10^{-1} and has an observed $C_ot_{1/2}$ of 0.003. The intermediate component reassociates next between C_ot values of 10^{-1} and 10^2 and has an observed $C_ot_{1/2}$ of 0.45. The slow component is the last fraction to reassociate with C_ot values ranging from 10^2 to 10^5 and has an observed $C_ot_{1/2}$ value of 3048.

A wealth of information can be deduced from C_ot curves:

- **The proportion of the genome in each class.** The fraction of the genome represented by each component is determined by extrapolating from the point where the reassociation is complete to the ordinate for each component. In *Figure 3.3* the fast, intermediate and slow components represent 14%, 30% and 56%, respectively, of the genome.
- **The real $C_ot_{1/2}$ of each component.** The observed $C_ot_{1/2}$ is not the 'real' $C_ot_{1/2}$ because the values of C_o used to calculate C_ot refer to the total DNA concentrations in the reassociation solution rather than

the concentration of each component. The 'real' $C_ot_{1/2}$ can be derived simply by multiplying the observed $C_ot_{1/2}$ by the gene fraction of the component. The fast component represents 14% of the DNA, so its concentration is $(0.14 \times$ total $C_o)$ and the 'real' $C_ot_{1/2}$ is $(0.14 \times$ observed $C_ot_{1/2})$ i.e. $0.14 \times 0.003 = 0.00042$. Similarly the 'real' $C_ot_{1/2}$ values of the intermediate and slow components are 0.135 and 1707, respectively.

- **The complexity of each component.** The slow component is assumed to be unique, so its complexity can be determined by reference to a standard of known complexity which has been reassociated under the same conditions. Suppose that *E. coli* DNA reassociates with a $C_ot_{1/2}$ of 4 under these conditions, then the complexity of the slow component can be deduced from the relationship:

$$\frac{C_ot_{1/2} \text{ (fast component DNA)}}{C_ot_{1/2} \text{ (E. coli DNA)}} = \frac{\text{Complexity of fast component}}{4.2 \times 10^6 \text{ bp}}$$

$$\text{Complexity of fast component} = \frac{0.00042 \times 4.2 \times 10^6}{4}$$

$$= 441 \text{ bp}$$

Similarly,

$$\text{Complexity of intermediate component} = \frac{0.135 \times 4.2 \times 10^6}{4}$$

$$= 1.42 \times 10^5$$

$$\text{Complexity of slow component} = \frac{1707 \times 4.2 \times 10^6}{4}$$

$$= 1.79 \times 10^9 \text{ bp}$$

- **The repetition frequency or relative number of copies.** The number of copies, R, of sequences in each component is inversely proportional to $t_{1/2}$ and hence to the observed $C_ot_{1/2}$ values. This follows from the fact that sequences present at higher concentrations reassociate faster than those at lower concentrations and hence have a lower $C_ot_{1/2}$ value. If the slowest reassociating fraction is present once per genome, then the number of copies of intermediate fraction can be deduced from the ratio of the $C_ot_{1/2}$ values:

$$R = \frac{\text{Observed } C_ot_{1/2} \text{ slow}}{\text{Observed } C_ot_{1/2} \text{ fast}} = \frac{3048}{0.003}$$

$$= 1.02 \times 10^6 \text{ copies}$$

Similarly, the repetition frequency of the intermediate component is 6773 copies

The results are summarized in *Table 3.1.*

Table 3.1. Summary of analysis of DNA reassociation

Component	Genome fraction	Observed $C_ot_{1/2}$	Real $C_ot_{1/2}$	Complexity if *E. coli* = 4	Repetition frequency
Fast	0.14	0.003	0.00042	441	1.02×10^6
Intermediate	0.3	0.45	0.135	1.42×10^5	6773
Slow	0.56	3048	1707	1.79×10^9	1

To obtain the number of base pairs per genome, X, the complexity of each component is multiplied by the number of times the component is repeated.

$$X = [(441) (1.02 \times 10^6)] + [(1.42 \times 10^5)(6773)] + [(1.79 \times 10^9)(1)]$$
$$= (4.5 \times 10^8) + (9.62 \times 10^8) + (1.79 \times 10^9)$$
$$= 3.2 \times 10^9 \text{ nucleotide pairs}$$

The molecular weight of the cellular DNA is obtained by multiplying the value of X by 660 (the average molecular weight of a nucleotide pair):

$$\text{Molecular weight} = 660 \times 3.2 \times 10^9$$
$$= 2.1 \times 10^{12}$$

If the molecular weight is determined by a different method such as chemical measurement and gives a value of say 4.0×10^{12}, this is approximately twice the value obtained above and would indicate that there were double the number of copies of all the components.

The fast, intermediate and slow reassociating components are usually referred to as highly repetitive, moderately repetitive and nonrepetitive (or unique) fractions of DNA, respectively.

The exact shape of C_ot curves and the proportion of DNA that falls into the highly repetitive, moderately repetitive and unique DNA classes differ from one organism to another. C_ot curves seldom show such discrete kinetic components as those in *Figure 3.3*. Instead the curves tend to be much smoother and it can be difficult to determine how many components are present. The data can be analyzed by computer using curve-fitting programs [1,4].

Note that for optical measurements, it takes a few minutes for the temperature to fall from the denaturation temperature to the incubation temperature. At typical DNA concentrations of $50 \, \mu\text{g ml}^{-1}$ this delay is equivalent to a C_ot of $0.01 \, \text{mol} \, l^{-1} \text{s}$ and during this time repetitive DNA may have reassociated. It is, therefore, not practical to measure reassociation of repetitive DNA by optical methods.

3.4.1 Determining the repetition frequency of an isolated sequence

The number of copies of any particular sequence can be analysed by a similar type of analysis. Tracer amounts of the DNA of interest (e.g. a cloned gene), are radioactively labeled and mixed with a vast excess of genomic DNA. After reassociation, the $C_o t_{1/2}$ values of genomic DNA and tracer DNA are determined. The repetition frequency of the gene is given by the ratio of the $C_o t_{1/2}$ values of the tracer and unique sequences calculated as in the worked examples above.

3.5 Isolation of different components

Reassociation of DNA in solution can be used to isolate components within a particular kinetic class. For example, to isolate fast reassociating DNA (i.e. repetitive DNA), denatured DNA is incubated as above until a $C_o t$ value is reached at which the repetitive fraction has reassociated, but the intermediate and slow components have not. The reaction mixture is fractionated on a hydroxyapatite column so that the single-stranded DNA flows through and the reassociated, repetitive DNA is retained on the column. The repetitive DNA is then eluted at a higher salt concentration.

Similarly, to isolate single-copy DNA, fragmented, denatured DNA is incubated in solution under conditions that permit double-strand formation. At a time when the fast and intermediate reassociating components are known to have reassociated, the incubation mixture is applied to a hydroxyapatite column. The single-stranded DNA which flows through the column is the low copy number or unique DNA fraction.

3.6 Analysis of repetitive DNA

The results in *Table 3.1* do not mean that there are precisely 6773 sequences with a complexity of 1.42×10^5 belonging to the moderately repetitive class of DNA. There are probably many different moderately repeated sequences each with its own complexity and repetition frequency. What the $C_o t$ curve tells us is the average behavior of the class during reassociation. A similar argument holds for the fast-reassociating component.

3.6.1 *Examining the relatedness of sequences*

Repeated sequences are not necessarily identical. They usually have similar sequences that have diverged to different extents from a common ancestor. On reassociation of repetitive DNA, duplexes will be perfectly base-paired if the complementary strands come from exact copies of the same sequence. However, if the DNA strands come from related, but nonidentical family members, the duplexes will contain mismatches. The closer the sequences are related, the fewer mismatches the hybrids will contain.

The degree of relatedness can be detected by analyzing melting curves. Hybrids with mismatches have reduced thermal stability compared with perfectly-matched hybrids and the more mismatches that are present, the lower the T_m will be. The relatedness of sequences can be examined by measuring the T_m of reassociated hybrids.

This can be carried out in several ways:

- The simplest method involves carrying out a reassociation experiment and isolating the different kinetic components on hydroxyapatite columns. The reassociated DNA is transferred to a temperature-controlled cuvette in a spectrophotometer which is capable of providing continuous readout of u.v. absorption. The temperature is raised by 1°C every minute and the change in optical density at 260 nm is measured
- In the second method, reassociation is carried out to a C_0t value of about 20 by which time the moderately repetitive DNA will have reassociated. The DNA is applied to a hydroxyapatite column in 0.14 M phosphate buffer at 50°C and the single-stranded DNA is washed off. The stability of the bound DNA is analyzed by thermal elution from the column. Thus, the temperature of the column is raised by 5°C steps between 50°C and 100°C. As the double-stranded DNA melts, it loses affinity for the column and is eluted. It can be detected in the eluate by absorption at A260 nm or by radioactivity if the DNA is labeled. Poorly matched DNA elutes at a lower temperature than well-matched DNA because its T_m is lower.
- In the third method the melting of reassociated DNA is measured in the presence of tetraethylammonium chloride (TEACl). After reassociation, the DNA is lightly treated with nuclease S1 to remove single-stranded tails and excise loops. The thermal melting properties of the duplexes are determined in the presence of tetraethylammonium chloride [5]. This compound binds to A:T base pairs and prevents them from melting at lower temperatures than G:C pairs. The effect of base composition on melting temperature is neutralized so that T_m is dependent only on the length of the DNA. TEACl treatment gives

better measurement of the T_m of double-stranded DNA than hydroxyapatite measurements because the single-stranded regions have been removed. TEACl melting curves are shifted to lower temperatures and are steeper than those obtained with HAP.

Figure 3.4 shows the thermal melting curve of the repetitive component of reassociated DNA as analysed by change in optical density (method 1 above). The repetitive DNA not only melts at lower temperatures than the native DNA, but it melts over a much wider range of temperatures. This reflects the fact that repetitive DNA contains families of sequences whose members are not identical, but are related to each other in varying degrees. The sequences that melt at the lowest temperatures are most distantly-related and contain more mismatches than the more closely-related sequences which melt at higher temperatures. The difference in T_m between the native and repetitive DNA in *Figure 3.4* is 9°C. The degree of divergence can be estimated using the relationship that every 1–1.7% mismatching of bases in a duplex reduces the T_m by 1°C [6,7]. So the repetitive DNA contains sequences that have diverged by 9–15.3%.

Note that the A_{260} of repetitive DNA at low temperature is much higher than that of native DNA, indeed it is much closer to the absorbance of single-stranded DNA, and the total increase in absorbance (hyperchromicity) is 0.13. Single-stranded DNA has a hyperchromicity of 0.06 while double-stranded DNA has a hyperchromicity of about 0.27. Since hyperchromicity measures the extent of sequence matching, the low hyperchromicity of the hybrids is additional evidence that they contain a high proportion of unmatched nucleotides.

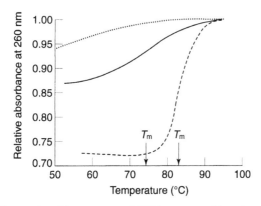

Figure 3.4. Thermal melting curves of DNA measured by absorbance at 260 nm. The melting curve of the isolated moderately repetitive component of DNA (—) compared to those of native (- - - -) and unique (single-stranded) (········) DNAs. The T_m s of native DNA and the repetitive fraction are 83°C and 74°C, respectively.

Hybridization conditions can be manipulated to control the formation of hybrids. At intermediate salt concentration and at temperatures close to T_m, only well-matched hybrids will form. Lower temperatures and higher salt concentrations permit the formation of less well-matched hybrids. By varying the conditions of hybridization, poorly matched duplexes can be selected against or allowed to form.

By varying the stringency of hybridization and studying the thermal melting behavior of reassociated DNA, valuable information can be obtained on mismatching within hybrids and thence on the relatedness of DNA sequences. It is not possible to give an absolute figure to the size of families because the size depends on the hybridization conditions used. The size of a family will appear larger if reaction conditions are used that permit distantly related repeated sequences to hybridize, whereas families will appear smaller if the conditions allow only well-matched hybrids to form.

3.6.2 Interspersion of unique and repetitive DNA

A single reassociation experiment gives no information on how repetitive and single-copy sequences are distributed within the genome. However, valuable information can sometimes be obtained by carrying out several reassociation experiments in which the fragment size of the DNA is varied.

Suppose that a single long molecule of DNA contains unique sequences interspersed with repeated sequences and that reassociation takes place under three different conditions.

- The DNA is not fragmented before denaturation. On renaturation, the repeated sequences will reassociate first and will carry with them the unique sequences into duplex molecules. So all the DNA will be scored as repetitive and there will be no separate unique component.
- The original DNA is broken into 'longish' fragments before denaturation. Some molecules will carry both unique and repetitive sequences on the same fragment, some will contain only repeated sequences and others will contain only unique sequences. On reassociation, the repeated sequences will again reassociate first. The fast reassociating fraction will contain two types of molecule: those that contain only repeated sequences and those that contain both repeated and unique sequences on the same fragment. A slower reassociating fraction will contain only unique sequences. The size of the repeated component will be smaller than in the situation above.
- The original molecule is broken into small fragments of a few hundred base pairs in length. The molecules will probably be too small to contain both repeated and unique sequences on the same fragment. So on denaturation and reassociation, the fastest reassociating component will contain only repeated sequences and the slow component

will contain only unique fragments. The size of the repetitive fraction will be lower than either of the above situations. Conversely the size of the unique fraction will be larger than either of the above situations.

By carrying out reassociation experiments with different sizes of DNA fragments followed by analyses of the T_m of the duplex DNA, the pattern of interspersion of repeated and unique sequences can be investigated [8].

It follows from these considerations that the size of a component depends on the length of fragments into which the DNA has been broken.

For a single nucleic acid species reassociating with itself, the hybrid melts over a very narrow temperature range and the T_m is the same irrespective of the temperature of reassociation, T_i. However, if the same nucleic acid is incubated with a complex mixture of sequences which have varying degrees of homology, the T_m profile depends on the hybridization conditions. For hybrids formed at low criterion (T_m $-25°C$), the melting temperature is broad because both well-matched and poorly matched hybrids are formed. They melt at different temperatures, so the overall melting profile which is a composite of the contributions of all the hybrids, will reflect this. At high criterion (T_m $-8°C$), only hybrids with a high degree of homology form so they melt over a very narrow temperature range. The temperature profile is also broad when variable length fragments are used. This is most apparent at short average lengths of DNA in accordance with the empirical relationship:

$$T_n - T_m = 650/L$$

where L is the length of the probe in nucleotides, T_m is the melting temperature of the short hybrid and T_n is the melting temperature of long DNA molecules [1].

References

1. **Britten, R.J., Graham, D.E. and Neufield, B.R.** (1974) *Methods Enzymol.* **29:** 363–418.
2. **Wetmur, J.G. and Davidson, N.** (1968) *J. Mol. Biol.* **31:** 349–370.
3. **Britten, R.J. and Kohne, D.E.** (1968) *Science* **161:** 529–540.
4. **Young, B.D. and Anderson, M.L.M.** (1985) In: *Nucleic Acid Hybridisation: A Practical Approach* (eds B.D. Hames and S.J. Higgins). IRL Press, Oxford, pp. 225–231.
5. **Powell, J.R. and Caccone, A.** (1990) *J. Mol. Evol.* **30:** 267–272.
6. **Bonner, T.I., Brenner, D.J., Neufield, B.R. and Britten, R.J.** (1973) *J. Mol. Biol.* **81:** 123–135.
7. **Caccone, A., DeSalle, R. and Powell, J.R.** (1988) *J. Mol. Evol.* **27:** 212–216.
8. **Graham, D.E., Neufield, B.R., Davidson, E.H. and Britten, R.J.** (1974) *Cell* **1:** 127–136.

4 Solution hybridization: RNA:DNA hybridization

4.1 Introduction

Information on the relationship between DNA and the transcripts it encodes can be obtained from RNA:DNA hybridizations. However, there are several factors which combine to make DNA:RNA hybridizations complicated – certainly more complicated than DNA:DNA reassociations.

- The complexity of RNA populations is much lower than that of DNA because only a small part of the genome is transcribed in any cell type.
- In most cell types not only are there many different transcripts, but these are present in a range of copy numbers stretching from very few to very many.
- In DNA:RNA hybridizations two competing reactions occur:

  DNA + DNA → DNA:DNA
  DNA + RNA → DNA:RNA

- In DNA:DNA reassociations, each fragment of single-stranded DNA has a complement, so at the end of reaction all the DNA should be double-stranded.

  However, in DNA:RNA hybridizations, if the input concentrations of DNA and RNA are similar, some sequences will be present in RNA excess (i.e. the concentration of some RNA species will exceed that of the DNA from which they were derived) whereas others will be in DNA excess. So some DNA will not enter hybrids and likewise some RNA will remain single-stranded at the end of the reaction. The kinetics of reaction are very complex and results are difficult to interpret.

Kinetics are much simpler to follow if either the RNA or DNA is present in large excess (Appendix B). These two experimental approaches provide different, but complementary information. If hybridization is carried out with DNA in vast excess, it is possible to determine if the RNA is derived from the repetitive, intermediate repetitive or unique component of DNA. In hybridizations with RNA excess information can be obtained on the complexity of RNA and on the number of copies of transcripts.

Experimentally, DNA:RNA hybridizations involve the following steps:

1. DNA is broken into small fragments of several hundred nucleotides in length by shearing, sonication, high speed blending or by digestion with a restriction endonuclease that cuts DNA frequently.
2. DNA and RNA are mixed in appropriate concentrations with the nucleic acid present in tracer amounts being radioactively labeled to make detection of hybrids easier.
3. Nucleic acids are denatured by heating briefly at about 90°C.
4. Incubation is carried out under conditions that facilitate hybridization (see Chapter 5).
5. The extent of hybridization (i.e. the fraction of the input single-stranded DNA and RNA that has become double-stranded) is measured at intervals.
6. The extent of hybridization is plotted against $C_o t$ or $R_o t$ where R_o is the concentration of single-stranded RNA at the start of incubation.

It may be more convenient to study the properties of RNA by synthesizing a cDNA copy followed by analysing hybridization of the cDNA.

4.1.1 Progress of reaction

There are several methods for following the progress of reaction.

Hydroxyapatite chromatography. As described in Section 3.2, hydroxyapatite chromatography can be used to separate single-stranded from double-stranded DNA. By virtue of its intrastrand secondary structure, RNA often binds to hydroxyapatite under conditions where single-stranded DNA does not. However, in the presence of formamide, urea or high concentrations of salt, single-stranded RNA does not bind to hydroxyapatite whereas RNA in a duplex does. So the progress of hybridization can be followed by applying the reaction mixture to a hydroxyapatite column in the presence of one of these compounds and measuring the proportion of DNA or RNA that is retained.

Nuclease resistance. Nuclease S1 is used extensively to monitor DNA:RNA hybridization under RNA excess conditions. The rationale is the same as described in Section 3.2.

Under appropriate conditions, single-stranded RNA is susceptible to digestion by pancreatic ribonuclease whereas RNA present in DNA:RNA hybrids is not. Samples containing labeled RNA can be digested to completion and analyzed for nuclease-resistant RNA.

Retention on nitrocellulose filters. Under appropriate conditions single-stranded RNA passes through a nitrocellulose filter whereas DNA and RNA in a duplex do not. The progress of hybridization can be followed by filtering aliquots of the reaction mixture and measuring the RNA-containing duplexes on the filter. Single-stranded RNA tails can be removed by treatment with pancreatic RNase prior to filtering.

4.2 DNA:RNA hybridizations with excess DNA

When DNA is in large excess and the RNA is present in minute quantities, the RNA sequences hybridize at the same rate as the DNA sequences from which they were transcribed (Appendix B). The rate depends only on the initial concentration, C_o, of DNA, so this type of reaction is said to be DNA-driven.

Under these conditions, the RNA tracer behaves just like the component of DNA from which it was derived (i.e. repetitive, moderately repetitive or nonrepetitive). That component can be identified by carrying out a DNA:RNA hybridization in vast DNA excess and following the rate at which both the DNA and the tracer enter duplexes. The $C_o t_{1/2}$ for RNA is used to determine the frequency class of DNA from which it was transcribed.

With a single RNA such as globin messenger RNA (mRNA), a simple curve with a single $C_o t_{1/2}$ value will be obtained. With a mixture of mRNA species such as polysomal RNA, a more complex curve will be obtained because each species will have its own $C_o t_{1/2}$ and the overall observed curve will represent the contribution of all the sequences present.

Results of a typical analysis of polysomal RNA are shown in *Figure 4.1*. About 70% of the mass of mRNA hybridizes at the same rate as the nonrepetitive component of DNA. The remainder hybridizes at the same rate as moderately repetitive DNA, but none with the same kinetics as highly repeated DNA. Generally up to about 80% of the hybridizing RNA is derived from single-copy DNA.

It must be emphasized that this type of analysis reveals nothing about:

- the number of unique DNA sequences represented in the mRNA;
- the sequence complexity of the mRNA;
- the number of copies of mRNA species.

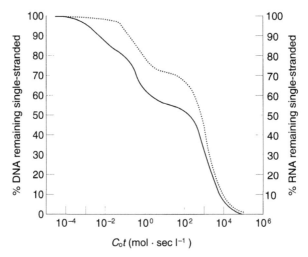

Figure 4.1. Hybridization of tracer polysomal mRNA (·········) to an excess of DNA (———). Most of the mRNA hybridizes at the same rate as the unique fraction of DNA. Some hybridizes at the same rate as the intermediate repetitive fraction, but none with the highly repetitive fraction.

This is because different genes are transcribed to different extents and since the RNA is measured in mass, there is not enough information here to determine how many copies of how many different transcripts there are. At one extreme the 70% of the mass of RNA that derives from single-copy DNA in *Figure 4.1* could come from a single gene while the 30% of the mass originating from the intermediate repetitive component could come from 20 000 or so genes. At the other extreme the distribution would be the converse, and of course many other possibilities exist.

More information on the number of genes transcribed and the number of copies of each transcript can be obtained by carrying out DNA:RNA hybridizations in RNA excess.

4.3 DNA:RNA hybridizations in RNA excess

If a small amount of DNA is hybridized to a large excess of mRNA, all the DNA complementary to the mRNA should be driven into hybrid provided that there is sufficient mRNA present to ensure that all the available DNA molecules have an RNA complement. The reaction is said to be RNA-driven and can be followed by measuring the amount of DNA that appears in hybrids as a function of R_0t (i.e. the initial RNA concentration × time). This type of analysis is referred to as R_0t analysis and the curve as an R_0t curve.

DNA:RNA hybridizations in RNA excess are used to measure the number of genes being expressed and the complexity of mRNA. To simplify interpretation of results, mRNA is usually hybridized to the nonrepetitive fraction of DNA that has been isolated on hydroxyapatite columns.

4.3.1 Estimation of the number of genes being expressed by saturation hybridization

mRNA present in vast excess (about 1000-fold excess by mass) is hybridized to fragmented, single-stranded, labeled DNA. Samples are removed at intervals and the fraction of DNA in the hybrid measured after nuclease S1 treatment. This ensures that all the hybrids being scored are base-paired along their whole length. The incubation is continued until no more DNA is driven into hybrid. By measuring the amount of DNA that is hybridized to the mRNA at the end of reaction, the proportion of DNA that is transcribed into mRNA can be determined. From this information an estimation can be made of the number of genes that have been transcribed.

This type of experiment is called a saturation experiment. What matters is the proportion of DNA that has become duplex at completion of hybridization, not the kinetics of the reaction. It is very important to ensure that incubation is continued until the slowest hybridizing species (i.e. rare species of RNA) have entered duplexes. This means that incubation to very high R_0t values is required. In addition, it is important that the reaction has reached completion because all the molecules which have the potential to hybridize do actually hybridize. If hybridization stops because either the DNA or RNA has been degraded, the calculations will over- or underestimate the proportion of DNA in hybrids at saturation.

Results of a saturation hybridization experiment in which single-copy DNA in tracer amounts is hybridized to a large excess of polyadenylated mRNA are shown in *Figure 4.2*. In *Figure 4.2* (a), results are expressed with a log scale for R_0t on the abscissa, and in *Figure 4.2* (b), the same results are re-plotted with a linear scale on the abscissa which allows the saturation level of DNA hybrids to be seen more clearly. The reaction is over by a R_0t value of 400 and, at saturation, 1.3% of the available DNA sequences have hybridized. Since transcription of structural genes is asymmetric, the 1.3% of DNA that has hybridized represents 2.6% of the complexity of nonrepetitive DNA.

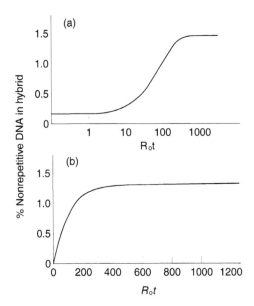

Figure 4.2. Saturation hybridization. Labeled, tracer, nonrepetitive DNA is hybridized to a vast excess of mRNA. The fraction of DNA hybridizing is plotted against R_ot. (a) R_ot is plotted on a log scale. (b) R_ot is plotted on a linear scale. A very small proportion of DNA appears in hybrids.

Suppose that the mRNA comes from the species analysed in Section 3.2 where the complexity of the nonrepetitive component of the DNA is 1.79×10^9 nucleotides (*Table 3.1*), then the complexity of the DNA represented in the polyadenylated mRNA population is $0.026 \times 1.79 \times 10^9 = 4.7 \times 10^7$ nucleotides.

If the average size of a mRNA molecule is 2000 nucleotides, then the total number of different polyadenylated messengers is $4.7 \times 10^7/2000 = 2.3 \times 10^4$. This represents the number of DNA genes that have been transcribed in the mRNA population under study.

4.3.2 Estimation of complexity of an RNA population by kinetic analysis of RNA-excess hybridization

Kinetics of DNA:RNA hybridizations are difficult to analyze because there is such a wide range of relative concentrations of DNA and RNA sequences. This complication can be avoided if mRNA is hybridized to its complementary DNA (cDNA) copy. Under these circumstances each mRNA sequence is represented in the cDNA at a frequency that reflects its occurrence in the mRNA, and since each cDNA molecule is derived from mRNA, all the cDNA should be driven into hybrid provided that there is sufficient excess of mRNA. So a 1000:1 mass excess of mRNA to cDNA reflects a 1000:1 excess at the sequence level.

In an RNA-excess hybridization, also known as a RNA-driven hybridization, with cDNA the following reaction takes place:

$$cDNA + RNA \rightarrow cDNA:RNA$$

The concentration of RNA remains essentially unchanged by the small amount of DNA:RNA hybrid that is formed and so the reaction follows pseudo first-order kinetics (see Appendix B). When the reaction is 50% complete $C/C_o = 0.5$ and

$$R_o t_{1/2} = \frac{\log_e 2}{k}$$

Where C/C_o is the fraction of DNA remaining single-stranded, R_o is the RNA concentration and t is the time of incubation. Provided that the concentration of RNA is in large excess over that of DNA, $R_o t_{1/2}$ is a measure of the rate constant, k, and is a measure of the complexity of the RNA [1–3].

The progress of a hybridization reaction is followed by measuring the proportion of DNA that remains single-stranded (C/C_o) as a function of $R_o t$. A $R_o t$ curve for hybridization of globin cDNA to an excess of globin mRNA is shown in *Figure 4.3*. Note that the curve is not symmetrical. From 10% to 90% hybridization occurs over a range of 1.5 log $R_o t$ values and the whole reaction takes place over about 2.5 log $R_o t$ values. This is typical of a pseudo first-order reaction in which one of the reactants is in vast excess over the other. It contrasts with second-order

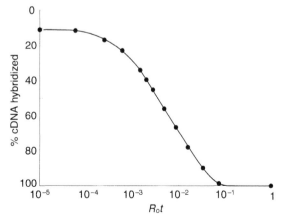

Figure 4.3. Time course of an ideal pseudo first-order reaction. Globin mRNA is hybridized to a vast excess of its cDNA.

reactions where the concentration of complementary strands is equal and the $C_o t$ range over which 10–90% hybridization occurs is 2 log units (*Figure 3.1b*).

Figure 4.4 shows a reassociation curve in which excess polyadenylated mRNA is hybridized to tracer amounts of its own cDNA. The reaction takes place over a range of about 3.5 log $R_o t$ values which is much larger than the 2.5 log $R_o t$ values taken by globin mRNA. These characteristics indicate that within the mRNA population sequences are present at varying frequencies [4]. Those hybridizing most rapidly are present at higher concentration than those hybridizing more slowly.

Just as a DNA reassociation curve is resolved into components in order to interpret it, so the $R_o t$ curve must be resolved into individual components. The curve in *Figure 4.4* is typical of mRNA:cDNA hybridizations in that it does not show discrete kinetic components and has to be resolved by computer using a curve-fitting program [5]. (This involves subdividing the overall curve into a number of components each with the characteristics of the pseudo first-order curve of globin RNA. The curve is analyzed into different numbers of components and these are summed to produce composite curves. The program works out which composite curve gives the best fit to the experimental data.) The computer resolves the overall curve in *Figure 4.4* into three components as shown by the dotted lines. These can be analyzed in terms of fraction of the population, $R_o t_{1/2}$, complexity and number of copies of each class.

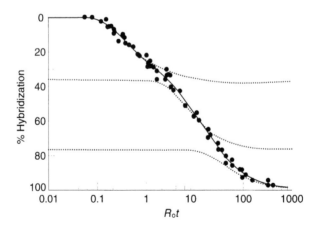

Figure 4.4. $R_o t$ curve for eukaryotic mRNA hybridizing to its own cDNA. The wide range of $R_o t$ values over which the reaction takes place indicates the presence of sequences at different abundances. The dotted lines show computer-generated components that could account for the experimental curve.

The first component to hybridize represents 23% of the cDNA and has an observed $R_o t_{1/2}$ value of 0.11. The real $R_o t_{1/2}$ value is $(0.23 \times 0.11) = 0.025$. By comparing the $R_o t_{1/2}$ value with that of a standard RNA whose complexity is known, the complexity of the component can be determined. Under the same experimental conditions globin RNA hybridizes with a $R_o t_{1/2}$ value of 5.8×10^{-4} mol s^{-1}. Globin mRNA consists of two different sequences, α and β, each being about 643 nt long. So its complexity is 1286 nt. Using this information, the complexity of the fast component can be derived from the relationship:

$$\frac{R_o t_{1/2} \text{ (fast component)}}{R_o t_{1/2} \text{ (globin RNA)}} = \frac{\text{Complexity of fast component}}{1286\,\text{nt}}$$

For the fast component , the complexity is

$$\frac{0.025 \times 1286}{5.8 \times 10^{-4}} = 5.5 \times 10^4 \text{ nt}$$

If the average complexity of an mRNA is known, the number of different species present in each component can be deduced. Assuming the average complexity of mRNA is 7×10^5 Da or 2000 nt, the number of different species in the fast component is $5.5 \times 10^4/2000 = 28$.

Abundance. The average number of copies of each mRNA species in a cell is known as its abundance or prevalence. The abundance of each component can be determined if the polyadenylated mRNA content of the cell is known. The mass of each component = (abundance $\times$ complexity)

There is typically about 0.32 pg polyadenylated mRNA per cell which is equivalent to

$$\frac{0.32 \times 10^{-12} \times 6 \times 10^{23}}{350} = 5.5 \times 10^8 \text{ nt}$$

where 6×10^{23} is Avogadro's number and 350 is the average molecular weight of a ribonucleotide.

The abundance of each component = fraction of total mass/complexity. For the fast component, this is $0.23 \times 5.5 \times 10^8/5.5 \times 10^4 = 2300$ nt

Analyses of the intermediate and slow components are carried out in the same way.

Results for all components are summarized in *Table 4.1*

The results in *Table 4.1* show that the fast component which makes up about a quarter of the mass of mRNA consists of a mere 28 species each of which is present in 2300 copies on average. Most of the complexity of

Table 4.1. Summary of analysis of cDNA:mRNA hybridization

Component	Fraction	Observed $R_0t_{1/2}$	'Real' $R_0t_{1/2}$	Complexity (nt)	No. of different species	Abundance
Fast	0.23	0.11	0.025	5.5×10^4	28	2300
Medium	0.4	1.0	0.4	8.9×10^5	445	247
Slow	0.37	16	5.92	1.31×10^7	6550	16

the mRNA is provided by 6550 different mRNAs which are present in only about 16 copies each.

4.3.3 Detection of sequences differentially expressed in two cell types

There are several methods by which the RNAs expressed in two cell types can be compared. Below two slightly different approaches based on saturation hybridization are described.

In the first method mRNA from the two types of cell is hybridized separately to tracer DNA in RNA driven reactions. In a third reaction, tracer DNA is hybridized to a mixture of the mRNAs [6].

Figure 4.4 shows a typical result. The mRNA from cell type A saturates 1.6% of single-copy DNA while mRNA from cell type B saturates 2.1%. If each cell type transcribes an entirely different set of genes, then hybridizing a mixture of the mRNAs should give an additive level of hybridization of 1.6 + 2.1 = 3.7% of the DNA. If all the mRNAs expressed in cell type A are also expressed in cell type B, the level of saturation should not be higher than the 2.1% shown by type B alone. In practice, the saturation level in the presence of equal amounts of mRNA from each cell type is 2.5% which is an increase of 16% (= 100 × [2.5 − 2.1]/ 2.5) of that observed with cell type B alone. This indicates that there is an 84% overlap of mRNA species between these two cell types. In mammals in which each cell type expresses 10 000–30 000 genes, this would mean that between 8400 and 25 200 mRNAs are expressed in common.

The second method also compares the mRNA population expressed in two different tissues by saturation hybridization, but gives more information on the degree of overlap. Suppose the mRNAs come from liver and kidney. Single-copy DNA is hybridized to saturation with excess mRNA from liver. The unreacted single-stranded RNA is removed by ribonuclease treatment and the DNA which hybridized to the liver mRNA is recovered. It is denatured and hybridized to an excess of mRNA from the kidney. The fraction of DNA that now hybridizes is a measure of the mRNA present in liver that is also expressed in kidney.

The fraction of DNA that did not hybridize to liver mRNA is recovered and hybridized to excess kidney mRNA. The proportion that now hybridizes is a measure of mRNA in the kidney that is not present in liver mRNA.

For completeness, the reciprocal experiment is also carried out (i.e. DNA that does and does not hybridize to kidney mRNA is recovered and hybridized to liver mRNA).

Saturation hybridization with sequential addition of mRNAs. A modification of the saturation hybridization technique is sometimes used to determine if two populations of cells contain different mRNAs [7]. *Figure 4.5* shows the results of an experiment designed to determine if the polyadenylated and nonpolyadenylated mRNAs in rat brain are transcribed from different sequences. The unique fraction of DNA is first hybridized to polyadenylated mRNA. At saturation, 4.5% of DNA has hybridized. An excess of poly(A)$^-$ mRNA is now added to the hybridization mixture, and a further 4.5% of the DNA is driven into hybrid. In a separate experiment total polysomal mRNA saturates 9.2% of DNA. These results show that in brain cells, the polyadenylated and nonpolyadenylated mRNAs are different sequences. Had they been the same, the addition of nonpolyadenylated mRNA would not have changed the amount of DNA appearing in hybrids.

Differential expression of mRNA sequences analyzed by kinetics of RNA-excess hybridizations. More detailed information on the overlap between mRNA in different cell types can be obtained by studying kinetics of hybridization. R_ot curves can be used to compare the mRNA populations from two or more cell types [8]. The mRNA from one cell type (e.g. liver) is hybridized to its own cDNA. It is also cross-hybridized to cDNA prepared from mRNA of another tissue such as

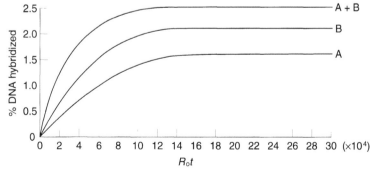

Figure 4.5. Differential expression of mRNA in two cell types analysed by saturation hybridization. Excess mRNA from cell type A saturates 1.6% of single-copy DNA. Excess mRNA from cell type B saturates 2.1%. When both mRNAs are present, 2.5% of single-copy DNA hybridizes.

brain or kidney. Analyses of $R_o t$ curves from such experiments show that there is about 70% cross-hybridization between mRNA of one tissue and cDNA of another. However, in cross-hybridizations, the fastest hybridizing component is missing. This means that the mRNAs that are most abundant in one tissue are not most abundant in the others.

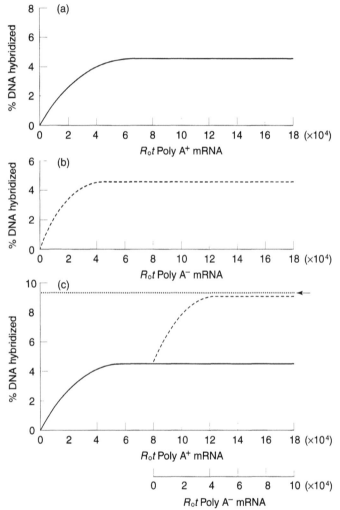

Figure 4.6. Successive hybridization of rodent brain mRNA fractions to single-copy DNA identifies different sequences in each fraction. (a) Excess rodent brain poly (A)$^+$ mRNA is hybridized to unique DNA. At saturation, 4.5% of the DNA is in hybrids. (b) With poly (A)$^-$ mRNA, 4.5% unique DNA is in hybrids. (c) Hybridization of poly (A)$^+$ mRNA to unique DNA takes place until a $R_o t$ value of 8×10^4 is reached when an excess of brain poly (A)$^-$ mRNA is added and incubation continued. The lower scale on the abscissa shows the $C_o t$ values for poly (A)$^-$ hybridization. The arrow shows the 9.2% saturation level obtained when unique DNA hybridizes with total polysomal RNA.

To investigate the overlap in expression of mRNAs further, the cDNA of one tissue is fractionated into three kinetic components: fast, intermediate and slow hybridizing fractions. These are dissociated and the cDNA from each is recovered and hybridized to mRNA from a different cell type. These experiments show that the mRNAs that are abundant in one tissue are expressed at lower levels in other tissues. Similarly, results show that there is extensive overlap in the middle frequency cDNAs from the three tissues.

References

1. **Birnsteil, M.L., Sells, B.H. and Purdom, I.F.** (1972) *J. Mol. Biol.* **63:** 21–39.
2. **Young, B.D. and Paul, J.P.** (1973) *Biochem. J.* **135:** 573–576.
3. **Hell, A., Young, B.D. and Birnie, G.D.** (1976) *Biochim. Biophys. Acta* **442:** 37–49.
4. **Bishop, J.O., Morton, J.G., Rosbach, M. and Richardson, M.** (1974) *Nature* **250:** 199–204.
5. **Young, B.D. and Anderson, M.L.M.** (1985) In: *Nucleic Acid Hybridisation: A Practical Approach* (eds B.D. Hames and S.J. Higgins). IRL Press, Oxford, pp. 225–231.
6. **Galau, G.A., Britten, R.J. and Davidson, E.H.** (1974) *Cell* **2:** 9–20.
7. **Axel, R., Felgelson, P and Schutz, G.** (1976) *Cell* **7:** 247–254.
8. **Hastie, N.D. and Bishop, J.O.** (1976) *Cell* **9:** 761–774.

5 Solution hybridization: reaction conditions

Reaction conditions affect the rate of duplex formation and thus the time for which incubation must be carried out in order that the reaction goes to completion.

5.1 The rate of formation of duplexes

The main variables that affect the rate of formation of duplexes are temperature, salt concentration and fragment length. In practice, laboratories carry out hybridizations under widely different conditions. So, in order to be able to compare results it is necessary to apply correction factors so that rates can be expressed under standard conditions. This entails knowing how the reaction conditions affect the rate. Standard conditions are $0.18\,M$ Na^+ ($0.12\,M$ sodium phosphate buffer, pH 6.8) at $60°C$ [1].

5.1.1 Temperature

The incubation temperature (T_i), affects the rate of formation of duplexes. *Figure 5.1* shows a typical temperature-dependence curve. Note that temperatures are usually expressed in terms of T_m, the melting temperature of the duplex. At low temperatures, the rate of reassociation proceeds very slowly. As the temperature is raised, the rate of reaction increases markedly until it reaches a broad maximum at approximately $15-30°C$ below the melting temperature of the DNA [2]. As T_i approaches closer to T_m the formation of duplexes competes with their melting so the apparent rate of reassociation falls. Reassociations/ hybridizations are often carried out at a T_i that is $20-25°C$ below the T_m in order to achieve the maximum rate of hybridization.

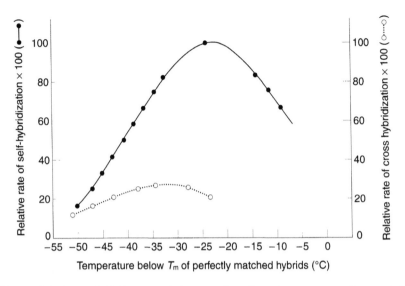

Figure 5.1. Rate of DNA reassociation as a function of temperature. Reassociation of perfectly-matched bacteriophage T4 DNA (●—●) and DNA with mismatches (○·····○). The temperature is expressed in terms of the T_m of perfectly matched DNA.

The rate of formation of hybrids containing mismatched bases also follows a bell-shaped temperature-dependence curve, but the maximum rate occurs at a lower temperature than for perfectly matched hybrids and the maximal rate is also lower. The rate of hybridization decreases about twofold for each 10°C reduction in T_m which is the equivalent of twofold for every 10% of mismatching [2]. For studying relatedness of sequences, hybrid formation is usually carried out at 25°C below the T_m for mismatched sequences.

For RNA:DNA hybrid formation, the temperature–rate response curve is also bell-shaped, but the maximum rate is 10–15°C below the T_m of hybrids [3]. Consequently, hybridizations are generally carried out at 10–15°C below the T_m.

5.1.2 Salt concentration

At low ionic strength, the rate of DNA:DNA hybridization is very low, but increases markedly as the salt concentration increases up to about 1.2 M NaCl when the rate becomes constant [4]. A very useful table detailing relative rates of DNA:DNA association at different ionic strengths has been drawn up [1]. For DNA:RNA hybridizations, the effect of salt concentration on the reaction rate has not been studied in such detail. However, the rate increases by 5–6-fold when ionic strength is increased from 0.2 to 1.5 M NaCl.

5.1.3 Fragment length

In DNA reassociation experiments where the length of both complementary strands is the same, the rate of reassociation is proportional to
the square root of the length of DNA [5]. This holds over three orders of
magnitude starting with $L = 100$ nt. If the reacting strands are of
different lengths, the rate of reaction is proportional to the square root of
the length of the shortest fragment [5]. In practice, fragments of about
300–500 nt are usually used, but for applications such as measuring the
interspersion of repeated sequences, longer fragments are used.

The relationship between length of fragment (L) and rate is:

$$k = 3 \times 10^5 \times L^{0.5} \times N^{-1}$$

where k is the rate constant and N the complexity. This equation holds
when T_i is optimal at $(T_m - 25)°C$ and the concentration of Na^+ is 1 M.

Experimentally, the size of DNA is reduced by physical methods such as
shearing, sonication and high-speed blending. For each preparation the
size is checked by standard methods such as denaturing gel electrophoresis. DNA can also be fragmented by digestion with a restriction
enzyme. The enzyme should be a frequent cutter or the width of the size
distribution curve will be great and the different sizes of fragment will
hybridize at very different rates.

5.1.4 Other factors

As discussed in Sections 3.3 and 4.3.2, the rate of duplex formation is
inversely proportional to the complexity for both DNA:DNA and
DNA:RNA hybridizations. Within the pH range 5–9, the renaturation
rate is essentially independent of pH [4]. There is an effect of base
composition on the rate of reassociation, but the effect is small and is
usually ignored. Increase in viscosity decreases the rate of reaction and
the effect can be quite large [4]. In practice, standard and unknown
DNAs are incubated under precisely the same conditions so that
viscosity affects both equally.

5.2 Preventing thermal breakdown of nucleic acid

Many reassociation experiments require nucleic acid to be incubated at
high temperatures (60–75°C) for several days. This creates problems
because prolonged incubation at elevated temperature causes the

nucleic acid to breakdown – thermal scission. Two main approaches can be taken to minimize this effect:

- speeding up the reaction,
- lowering the incubation temperature without altering the stringency.

5.2.1 *Speeding up the reaction*

There are several useful means of increasing the rate of reaction.

- *Increasing the concentration of nucleic acid.* The extent to which this method can be used is limited by the solubility of DNA and RNA.
- *Increasing the salt concentration.* This is a very useful means of increasing the rate without altering the concentration of DNA. As noted above the relative rates of hybridization have been evaluated for different concentrations of sodium phosphate and sodium chloride which makes it easy to convert C_0t values into standard salt conditions [1].
- Several compounds accelerate the rate of hybridization by effectively concentrating the nucleic acid in the reaction mixture. Inert, high-molecular-weight polymers such as polyethylene glycol and dextran sulfate form two phase systems in solution and exclude the nucleic acid from the space they themselves occupy. This concentrates the nucleic acid in solution [6,7].

The rate of reaction can also be increased markedly by incubation in the presence of phenol [8]. The reaction is thought to take place at the phenol/aqueous interface which causes a high local concentration of DNA. The rate is increased most effectively, some 25 000-fold, if the concentration of nucleic acid is low $(4\,\mu g\,ml^{-1})$. At higher concentrations of DNA $(1.6\,mg\,ml^{-1})$ the increase drops to about 35-fold probably because the interface has become saturated. The rate and extent of hybrid formation in the presence of phenol are hard to control and the procedure has not been extensively used.

5.2.2 *Lowering the incubation temperature*

A widely used method for preventing thermal breakdown of nucleic acids is to add a denaturing agent and lower the incubation temperature. Formamide destabilizes nucleic acid and reduces the T_m by between 0.5 and 0.75°C for each 1% increase in formamide concentration. The exact amount by which the T_m is lowered depends on the (G+C) content of the DNA [9], but an average value of about 0.63°C for each 1% increase in formamide concentration is generally accepted. In the presence of formamide, the rate of DNA:DNA reassociation maintains a bell-shaped temperature-dependence curve,

but the optimal rate is reduced by about 1.1% for every 1% formamide
[10]. DNA:DNA hybridizations in the presence of formamide are usually
carried out at 25°C below the T_m in formamide.

The depression of the T_m of DNA:RNA and RNA:RNA hybrids by
formamide is nonlinear. In high concentrations of formamide, DNA:
RNA hybrids are formed in preference to DNA:DNA hybrids. In 80%
formamide the rate of DNA:DNA hybridization is three-fold lower than
in aqueous solution and 12-fold lower for DNA:RNA hybridization [9].

Although formamide is probably the most extensively used reagent for
lowering the T_m, other denaturants are also useful. These include urea,
trimethylammonium chloride and triethylammonium chloride, ethylene
glycol and sodium perchlorate [11].

5.3 Factors affecting the stability of hybrids

The stability of hybrids depends on the reaction conditions. This
property is exploited to select for or against particular hybrids on the
basis of their stability.

The T_m is a measure of the stability of hybrids. It is affected by ionic
strength, base composition, length of fragment to which the nucleic acid
has been reduced, the degree of mismatching and the presence of
denaturing agents. The following relationship shows how the T_m is
affected by the various components of the incubation mixture. It
combines results from several sources.

For DNA:DNA hybrids

$$T_m = 81.5°C + 16.6 \log_{10} \left(\frac{[Na^+]}{1+0.7[Na^+]} \right) + 0.41 \ (\% \ G+C) - 500/L$$
$$- \ P - 0.63 \ (\% \ formamide) \tag{5.1}$$

where Na^+ is the monovalent cation concentration, (% G+C) is the
percentage of G and C nucleotides in the DNA and L is the length of the
duplex in bp. P is the percent mismatching (discussed below). The
relationship holds for DNAs with (G+C) contents of 30–75%.

For DNA:RNA hybrids:

$$T_m = 67°C + 16.6 \log_{10} \left(\frac{[Na^+]}{1+0.7[Na^+]} \right) + 0.8 \ (\% \ G+C) - 500/L$$
$$- \ P - 0.5 \ (\% \ formamide) \tag{5.2}$$

For RNA:RNA hybrids:

$$T_m = 78°C + 16.6 \log_{10} \left(\frac{[Na^+]}{1+0.7[Na^+]} \right) + 0.7 \, (\% \, G{+}C) - 500/L$$
$$- P - 0.35 \, (\% \, \text{formamide}) \hspace{3cm} (5.3)$$

Formamide depresses the T_m of DNA:DNA hybrids by about 0.63°C for every 1% increase in formamide concentration. The melting temperatures of RNA:RNA and RNA:DNA hybrids are also reduced in formamide-containing solutions, but the T_m is lowered by less than that for DNA:DNA hybrids and the response is not linear with formamide concentration [12].

In aqueous solutions the stability of DNA:RNA hybrids is greater than that of DNA:DNA duplexes of the same sequence and the relative stability is enhanced in the presence of high concentrations of formamide. This property can be exploited to allow RNA:DNA hybrids to form selectively in the presence of complementary strands of DNA. Incubation at 70% formamide and 41–50°C completely suppresses competing reassociation of DNA while allowing complete formation of DNA:RNA hybrids [9,13].

5.3.1 Mismatches

Nucleic acids containing mismatched bases are less stable thermally than perfectly matched duplexes. The more mismatches there are, the less stable hybrids are. Every 1.7% base pair mismatching in DNA reduces the T_m by 1°C [14].

Reassociation experiments often contain mixtures of sequences that are related to each other to different extents. If incubation is carried out close to the T_m of perfectly base-paired duplexes, mismatched hybrids will not be stable and will be selected against. Conditions which prevent formation of mismatched sequences are said to be stringent or of high stringency. The difference between T_m and the incubation temperature, T_i is called the criterion. At small criteria, the incubation temperatures are close to the T'_m which selects for well-matched or closely related duplexes.

To detect distantly related sequences, conditions must be made favorable for mismatched hybrids to form. This means that the incubation temperature must be reduced considerably. These conditions are said to be relaxed, open or of low stringency. It should be noted that under conditions which permit poorly matched sequences to form, the rate of hybridization is less than that for perfectly matched hybrids and this needs to be taken into consideration when planning the total time of incubation.

References

1. **Britten, R. J., Graham, D.E. and Neufield, B.R.** (1974) *Methods Enzymol.* **29:** 363–418.
2. **Bonner, T.I., Brenner, D.J., Neufield, B.R. and Britten, R.J.** (1973) *J. Mol. Biol.* **81:** 123–135.
3. **Nygaard, A.P. and Hall, B.D.** (1964) *J. Mol. Biol.* **9:** 125–142.
4. **Wetmur, J. and Davidson, N.** (1968) *J. Mol. Biol.* **31:** 349–370.
5. **Wetmur, J.R.** (1971) *Biopolymers* **14:** 601–613.
6. **Chang, C.-T., Hain, T.C., Hutton, J.R. and Wetmur, J.G.** (1974) *Biopolymers* **13:** 1847–1858.
7. **Wetmur, J.R.** (1975) *Biopolymers* **14:** 2517–2524.
8. **Kohne, D.E., Levinson, S.A. and Byers, M.J.** (1977) *Biochemistry* **16:** 5329–5341.
9. **Casey, J. and Davidson, N.** (1977) *Nucleic Acids Res.* **4:** 1539–1552.
10. **Hutton, J.R.** (1977) *Nucleic Acids Res.* **4:** 3537–3555.
11. **Young, B.D. and Anderson, M.L.M.** (1985) In *Nucleic Acid Hybridisation: A Practical Approach.* (eds B.D. Hames and S.J. Higgins). IRL Press, Oxford, pp. 47–71.
12. **Wetmur, J.G.** (1991) *Crit. Rev. Biochem. Mol. Biol.* **4:** 227–259.
13. **Vogelstein, B. and Gillespie, D.** (1977) *Biochem. Biophys. Res. Commun.* **75:** 1127–1132.
14. **Caccone, A., DeSalle, R. and Powell, J.R.** (1988) *J. Mol. Evol.* **27:** 212–216.

6 Basic types of filter hybridization

In filter hybridization, single-stranded DNA or RNA is bound to an inert support in such a way that it is prevented from reacting with itself, but is available to form hybrids with single-stranded nucleic acid added in solution. The sequences in solution are known as the probe and are labeled with reporter molecules. After hybridization, nonreacting probe is removed by washing and hybrids on the filter are visualized by virtue of the reporter molecules. The filter is said to have been 'probed' for the sequence of interest, for example if the reacting species in solution is ribosomal RNA, the filter is said to have been probed for rRNA.

All applications of filter hybridization are variations of four basic methods which are described below.

6.1 Probing recombinant libraries

Filter hybridization is used to search libraries of recombinant bacteriophages, phages or plasmids for sequences of interest [1–3]. The recombinant phages are plated out on dishes and incubated under standard conditions so that phage plaques are produced in a background of unlysed bacteria (*Figure 6.1*). A sterile filter is placed on the surface of the plate. Some of the phages in each plaque stick to the filter so that when the filter is removed, it carries a replica of the pattern of plaques on the plate. The phages on the filter are lysed, and the DNA is made single-stranded by placing the filter on absorbent paper saturated with appropriate solutions. The filter is then neutralized and treated to bind the single strands firmly to the filter. Hybridization is carried out by incubating the filter with a single-stranded probe. The most frequently used probes are labeled with ^{32}P-nucleotides.

After hybridization, the probe is washed away and the filter is exposed to X-ray film. Where hybrids have formed, the decay of ^{32}P label in the

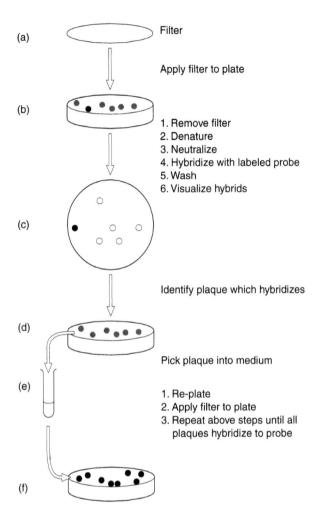

(a) Filter

Apply filter to plate

(b)

1. Remove filter
2. Denature
3. Neutralize
4. Hybridize with labeled probe
5. Wash
6. Visualize hybrids

(c)

Identify plaque which hybridizes

(d)

Pick plaque into medium

(e)

1. Re-plate
2. Apply filter to plate
3. Repeat above steps until all plaques hybridize to probe

(f)

Figure 6.1. Probing a recombinant phage library detects plaques which hybridize with the probe. Positively scoring plaques can be recovered from the plate and grown up.

hybridized probe exposes the film giving rise to a black dot. By aligning the autoradiograph with the filter and the original plate, the plaque hybridizing to the probe can be identified. Since only a small fraction of phage in a plaque is transferred to the filter, it is possible to excise the remainder of the plaque from the plate and rescue live phage. In the first plating out, the high density of phage makes it likely that the plaque picked will be contaminated by neighboring phage, so the procedure is repeated several times until all the phage on the plate hybridize with the probe and thus contain the desired cloned sequence. *Figure 6.2* shows results of hybridizing replicate plaque lifts with two different probes.

For probing recombinant libraries in plasmids, bacteria containing the plasmids are spread on a plate and incubated to allow colonies to grow up. The colonies are transferred to filters and probed in a similar way to that described for phage. This allows colonies hybridizing to the probe to be located and purified.

6.2 Southern blots

The technique of Southern blotting is named after its inventor, Ed Southern [4]. DNA that has been digested with restriction enzymes is applied to a gel (usually agarose) and separated according to size by electrophoresis. The gel is treated with alkali to denature the DNA and a filter is placed over the gel (*Figure 6.3*). The DNA is transferred to the filter by capillary diffusion. The pattern of fragments of DNA on the filter is a replica of that in the gel.

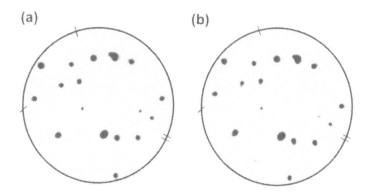

Figure 6.2. Detection of phage plaques containing rearranged human Ig lambda genes. The experiment was designed to isolate a rearranged human Ig lambda gene containing both C_λ and V_λ sequences (in most cell types these sequences are located far apart on chromosome 22 and would not normally be found on the same restriction fragment of DNA). A recombinant genomic library of bacteriophage λ containing fragments from an Ig-producing cell line was probed with an IgC$_\lambda$ sequence. One of the positive plaques was re-plated at low density and duplicate plaque lifts were probed with (a) the same C_λ probe and (b) a V_λ probe excised from a cDNA. Most plaques on the plate hybridized to both probes so the cloned sequence contained both C_λ and V_λ sequences and hence a re-arranged Ig gene.

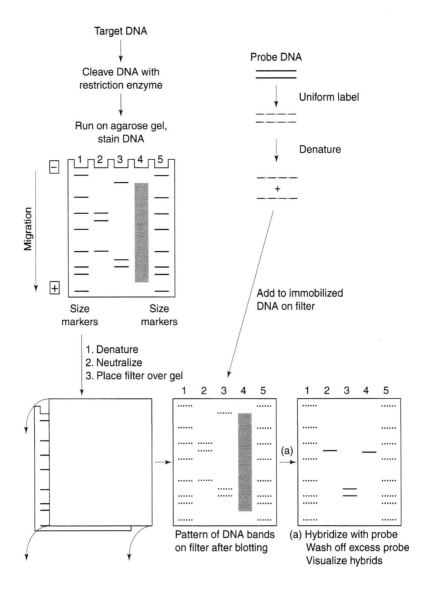

Figure 6.3. Southern blotting detects fragments of DNA that have been size fractionated by gel electrophoresis. Lanes 1 and 5: size markers; lanes 2 and 3: cloned DNA; lane 4: genomic DNA.

The DNA is firmly fixed to the filter and hybridized as described for phage probing. The position of the band (autoradiographic signal) is a measure of its size whereas the intensity of the band is a measure of the number of copies of sequences that hybridize.

6.3 Northern blots

Northern blots are similar to Southern blots except that the nucleic acid being analyzed is RNA instead of DNA [5]. The RNA is denatured before electrophoresis, so the gel does not need to be pretreated before transferring the RNA to the filter. The probes are often cloned genes, so that information is obtained on expression of genes and on the size and relative amounts of transcripts. The existence of splice variants may also be detected. *Figure 6.4* shows a Northern blot probed for rare mRNAs.

6.4 Dot blots and slot blots

In dot blots, DNA or RNA in solution is spotted on to filters and allowed to dry. For slot blots, the procedure is similar except that the nucleic acid is applied through a slot-shaped template. The nucleic acid is denatured and bound firmly to the filters. Hybridization is carried out

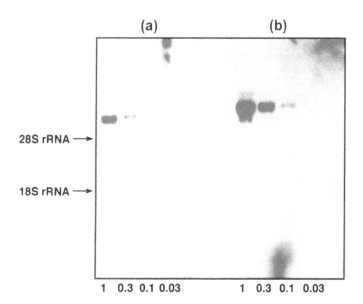

Figure 6.4. Probing rare mRNAs on a Northern blot with DIG-labeled RNA probes. Total human liver RNA was separated on a MOPS-formaldehyde gel and blotted on to a charged nylon filter. (a) Probed with DIG-labeled HMG Co-reductase antisense RNA. (b) Probed with DIG-labeled LDL-receptor antisense RNA. Hybridization in 50% formamide-containing buffer at 68°C overnight with 100 ng probe/ml. Filters were washed under stringent conditions and hybrids were detected by alkaline phosphatase action on AMPDD (see *Chapter 13*).

with labeled probes as described above (*Figure 6.5*). The autoradiograph shows which dots contain sequences complementary to the probe.

Dot blots are faster to set up than Southern or Northern blots because gels do not have to be run and the transfer of nucleic acid to the filter is far quicker. However, dot blots give no information on the size and number of different sequences contributing to the hybridization signal. Dot blots are well-suited to analysis of many samples at once and have the added advantage that it is easy to prepare replicate filters. This allows filter-bound sequences to be analyzed under different conditions, for example, with different probes or with different hybridization and washing conditions.

Figure 6.6 shows the results of probing replicate RNA dot blots with three different probes. The rDNA probe establishes that the dots contain equal amounts of RNA. The Ha-*ras* probe shows that the tumor tissue expresses more Ha-*ras* transcripts than normal tissue. Differential expression is specific for the Ha-*ras* oncogene because it is not shown by a c-*sis* probe.

Dot blot hybridization can be used qualitatively and is capable of great discrimination in that it can distinguish between closely related

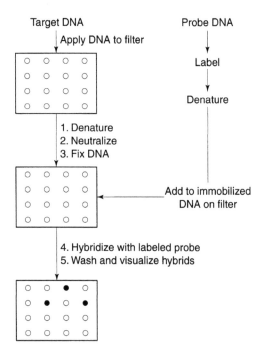

Figure 6.5. Dot blots detect sequences in unfractionated nucleic acid that has been bound to the filter.

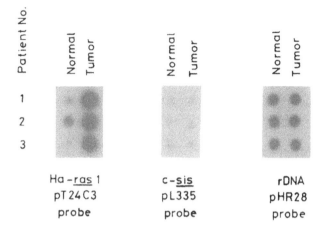

Figure 6.6. RNA dot blots. Replicate filters containing 10 μg total RNA from normal and tumor breast tissue from different patients were separately hybridized with Ha-*ras*-1, c-*sis* and rDNA probes. The autoradiograph of rDNA probe confirmed that equal amounts of RNA had been applied to the dots. In all patients the transcripts homologous to the Ha-*ras*-1 probe were more abundant in tumor than in normal tissue. The result was specific to the *ras* oncogene since the c-*sis* probe did not show significant hybridization to any sample.

members of multigene families and between sequences that differ by a single nucleotide. This property is exploited in the detection of mutations in prenatal diagnosis of genetic disease. The technique can also be used quantitatively with appropriate calibration, but is more commonly used as a semiquantitative method for estimating the relative levels of sequences in different samples.

These basic techniques are used to isolate recombinant clones, map genes, to study transcription, to distinguish between different members of related sequences and to detect single base changes in nucleic acid. The methods are very versatile and if used carefully, are robust and sensitive. Applications of these methods are discussed in Chapter 15.

Experimentally, all filter hybridization can be divided into the following steps (see *Figure 6.7*):

1. Binding single-stranded nucleic acid to the filter;
2. Prehybridizing the filter in order to coat nonspecific sites and minimize the background 'signal';
3. Preparation of the probe;
4. Incubating the filter with single-stranded probe under conditions that favor formation of hybrid;
5. Washing to remove nonhybridized probe and to dissociate mismatched hybrids;
6. Visualizing hybrids.

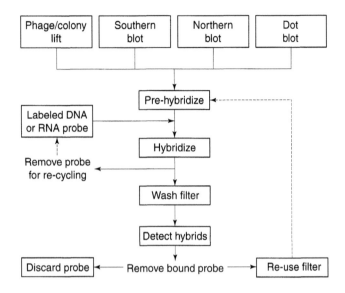

Figure 6.7. Flow diagram showing the steps involved in filter hybridization.

There are many different ways of carrying out each step and the choices made can determine which hybrids form and which persist. Hybridization experiments need to be well planned in order to get satisfactory results. Chapters 7–9 discuss hybridization strategy and choice of the probe. Chapters 10–13 give experimental details for some of the most commonly used procedures.

References

1. **Sambrook, J., Fritsch, E.F. and Maniatis, T.** (1989) In: *Molecular Cloning: A Laboratory Manual*, 2nd Edn. Cold Spring Harbor Laboratory Press, Cold Spring Harbor, NY.
2. **Grunstein, M. and Hogness, D.** (1975) *Proc. Natl Acad. Sci. USA* **72:** 3961–3965.
3. **Benton, W.D. and Davis, R.W.** (1977) *Science* **196:** 180–182.
4. **Southern, E.** (1975) *J. Mol. Biol.* **98:** 503–517.
5. **Davis, P.S.** (1980) *Proc. Natl Acad. Sci. USA* **77:** 5201–5205.

7 Hybridization strategy: long probes

7.1 Introduction

There is no single protocol that is appropriate for all applications of filter hybridization. Many choices are available at each step and these have to be carefully weighed up when designing an experiment. The hybridization strategy selected will depend on the purpose of the experiment and factors such as the materials available, the sensitivity required and whether or not related (mismatched) sequences are to be detected. The conditions of hybridization and washing have a profound influence on determining which hybrids form and which survive. As in solution hybridization, conditions under which only well-matched hybrids survive are termed stringent whereas conditions which allow mismatched hybrids to form and persist are termed open or relaxed.

This and the following chapter discuss the basic principles of filter hybridization so that the choice of reaction conditions can be based on an understanding of how they will affect hybridization. They also serve as a guide as to how conditions can be manipulated to optimize results and as a basis for troubleshooting.

7.2 Factors affecting the rate of hybridization and stability of hybrids

Many of the variables which affect the rate or extent of filter hybridization and the stability of hybrids have not been extensively studied. In the absence of direct data, it is common to use relationships that have been derived for solution hybridization, but these can only give approximate conditions for filter hybridization and optimization has to be carried out empirically.

No equations are derived here and the reader is referred to references [1,2] and references therein for more details.

7.2.1 Concentration of nucleic acids

Relationship between hybridization rate and nucleic acid concentration. The initial rate of filter hybridization depends on the concentrations of the probe in solution and the sequences bound to the filter (µg cm^{-2}). Kinetics differ depending on which is present in (molar) excess.

Filter hybridization depends on two processes: diffusion of the probe to the filter and hybridization. At low concentrations of filter-bound sequences, the rate-limiting factor is hybridization itself and the initial rate of hybridization depends on the concentrations of both the filter-bound and probe sequences. If the local concentration of filter-bound sequence is very high as in dot blots of plasmid DNA, hybridization may take place so quickly that the surrounding solution becomes depleted of probe and the overall reaction is limited by the rate of diffusion of the probe to the target on the filter [1].

Since many filter hybridizations seek to hybridize as much probe as possible to a molar excess of filter-bound sequences, this problem is important. The overall rate of hybridization will be speeded up by factors which increase diffusion of the probe to the target. These include using a small probe, small reaction volume, high temperature and shaking the reaction vessel.

If the probe is single-stranded and contains no self-complementary regions, the rate of hybridization should increase as the concentration of probe increases irrespective of whether the probe or filter-bound sequences are in excess. In practice, however, probe concentrations greater than about $100\,\text{ng}\,\text{ml}^{-1}$ are not used because they tend to generate high backgrounds. (Background refers to the nonspecific binding of probe to the filter. If background is high, it may be difficult to discern where the DNA has hybridized on the filter.)

Problems with high concentrations of double-stranded probe. If the probe is a simple double-stranded molecule, such as a restriction fragment of DNA, there will be two competing reactions: reassociation of the probe in solution and hybridization to target sequences on the filter. If the concentration of probe is high, reassociation will be favored over hybridization. This not only reduces the rate of hybridization, but also reduces the amount of probe that can hybridize to the filter. The effects of reassociation should not be ignored since as much as 20–30% of the double-stranded probe may be unavailable for hybridization to the filter [3].

Another problem may arise with probes of randomly sheared or tandemly repeated DNA. Reassociation in solution can lead to formation of duplexes with single-stranded tails which can then reassociate with other probe molecules to form bigger and bigger networks. The network eventually hybridizes via its tails to the target on the filter. The consequence is that as much as 10% of added denatured probe which apparently 'hybridizes' to the filter is of homologous rather than complementary sequence [3].

To avoid these complications conditions should be chosen that favor hybridization to the filter rather than reassociation in solution. This entails using a short probe, preferably single-stranded and at low concentration, and a small reaction volume to facilitate diffusion.

Determination if the probe or immobilized sequence is in excess. As noted above the kinetics of filter hybridization differ depending on whether the sequences in solution or bound to the filter are in excess. It is not always obvious which are in excess and a complicating factor is that some of the sequences bound to the filter seem to be unavailable for nucleation [1]. This means that although the amount of nucleic acid bound may be known, the amount that can participate in hybridization may not.

To check which is in excess, the following simple experiments can be performed.

- Prepare replicate dot blots and vary the input of probe. If filter-bound sequences are in excess, the hybridization signal will be proportional to the input of probe.
- Prepare several blots with varying amounts of bound nucleic acid and keep the input of probe constant. If probe sequences are in excess, the hybridization signal will be proportional to the amount of filter-bound nucleic acid. However, if the filter-bound sequences are in excess, the hybridization signal will be constant.

In practice, the amount of nucleic acid bound to the filter depends on the purpose of the experiment and the availability of the nucleic acid. Using 1–10 µg of mammalian DNA per lane of a Southern blot, single-copy genes can easily be detected with $20\,\mathrm{ng\,ml^{-1}}$ (typically 2×10^7 c.p.m. $\mathrm{ml^{-1}}$) ^{32}P-labeled probe. Northern blots commonly use 10–20 µg total RNA or 1–2 µg poly(A)$^+$ RNA per lane, but if the mRNA species being detected constitutes 0.01% or less of the polyadenylated RNA, the amount of poly(A)$^+$ RNA should be increased to about 10 µg per lane.

7.2.2 *Length of the probe*

For filter hybridization, the length of the probe can have two different effects. If the filter-bound sequences are in excess, the rate of hybridization decreases as the length of the probe increases. This occurs with both single-stranded and double-stranded probes. If the probe sequences are in excess, the rate is independent of the length of the probe [1].

In practice, DNA probes are usually about 400–500 bp in length. If they are longer than about 1500 bp, high backgrounds may occur. The size of DNA is usually controlled in the labeling procedure: in nick translation by the ratio of enzymes and in random priming by the ratio of primers to DNA. The length of RNA probes is less easy to control. For probes labeled by run-off transcription of recombinant DNA, an upper limit to the length can be made by the choice of the restriction enzyme used to linearize the plasmid (see Section 11.3.3). Incubation in mild alkali can then be used to reduce the length further.

7.2.3 Complexity of the probe

If probe sequences are in significant excess, the rate of hybridization is inversely proportional to the complexity of the probe. If hybridization is limited by diffusion of the probe to the filter, the rate is independent of the complexity of the probe [1].

7.2.4 Base composition

The rate of hybridization increases as the (G+C) content of the nucleic acid increases, but the effect is small and is usually ignored. A more important effect of base composition is the effect on the T_m of hybrids. The higher the (G+C) content, the greater is the stability of hybrids and the higher the T_m (Section 1.2.2). The (G+C) content of nucleic acids can be obtained from standard reference sources and T_m can be derived from relationships such as those in Section 7.2.6.

7.2.5 Salt concentration

Effect on hybridization rate. At low ionic strength, the rate of hybridization is low and increases markedly as the ionic strength increases [4]. However, above 0.4 M, the rate constant increases only twofold between 0.4 and 1 M Na^+ then reaches a plateau for DNA:DNA hybridization. Between 0.18 M and 1 M NaCl, the rate of RNA:DNA hybridization increases only twofold.

In practice DNA:DNA hybridizations are usually carried out at about 1 M NaCl, for example in 6 × SSC. For long hybridizations or when formamide is present, 6 × SSPE may be preferred because it has superior buffering capacity. (See Appendix D for composition of solutions.) For DNA:RNA and RNA:RNA hybridizations, reactions are usually carried out at 0.45 – 1 M NaCl: for example, in 2–5 × SSPE.

Effect on stability of hybrids. Ionic strength affects the stability of hybrids (see *Figure 1.4b*). High salt concentrations are needed to stabilize imperfectly matched duplexes. As noted above the salt concentration is usually kept high during hybridization, but can be manipulated during washing to stabilize or dissociate imperfectly matched hybrids according to the requirements of the experiment.

- Well-matched hybrids are washed at high stringency (low salt concentration, high temperature). For DNA:DNA hybrids the salt concentration may be reduced to 0.1 × SSC, whereas for RNA:RNA and DNA:RNA hybrids the salt concentration is generally 0.1–1 × SSPE. Under these conditions, poorly matched hybrids will dissociate.
- For preserving imperfectly matched hybrids, the salt concentration of the washing solution is usually raised to stabilize the hybrids. Washing is often carried out at 2–6 × SSC depending on the degree of homology in the hybrids, but optimal conditions have to be determined by trial and error.

7.2.6 Temperature

At low temperature the rate of hybridization is low, but increases as the temperature is raised until it reaches a broad maximum that is 20–30°C below the T_m of the hybrid [5]. The rate then declines as the temperature approaches T_m (see *Figure 5.1*). Similar temperature-dependence curves are found for imperfectly matched hybrids, but the maximum rate is displaced towards lower temperatures. For RNA:DNA hybrids similar curves are found [6] but the maximum rate is 10–15°C below the T_m.

Ideally, hybridization should be carried out at a temperature for which the rate is maximal, i.e. 20–25°C below the T_m for DNA:DNA hybridizations. However, the T_m is probably not known and is seldom measured. Instead T_m is estimated from relationships such as the following which are compiled from several sources [7,8].

For DNA:DNA hybrids:

$$T_m = 81.5°C + 16.6 \log_{10} \left(\frac{[Na^+]}{1+0.7[Na^+]} \right) + 0.41 \ (\% \ G+C) - 500/L$$
$$- P - 0.63 \ (\% \ \text{formamide}) \tag{7.1}$$

where Na^+ is the monovalent cation concentration, (% G+C) is the percentage of G and C nucleotides in the DNA and L is the length of the duplex in bp. P is the percent mismatching to be discussed below. The relationship holds for DNAs with (G+C) contents of 30–75%.

For DNA:RNA hybrids:

$$T_m = 67°C + 16.6 \log_{10} \left(\frac{[Na^+]}{1+0.7[Na^+]} \right) + 0.8 \, (\% \text{ G+C}) - 500/L$$
$$- P - 0.5 \, (\% \text{ formamide}) \tag{7.2}$$

For RNA:RNA hybrids:

$$T_m = 78°C + 16.6 \log_{10} \left(\frac{[Na^+]}{1+0.7[Na^+]} \right) + 0.7 \, (\% \text{ G+C}) - 500/L$$
$$- P - 0.35 \, (\% \text{ formamide}) \tag{7.3}$$

(Although formamide affects the T_m of both DNA:RNA and RNA:RNA hybrids, the depression is not linear with formamide concentration. The T_m probably lies somewhere between the values calculated by including and excluding these terms.)

These equations are derived from solution hybridization studies, but take into account the higher salt concentrations appropriate for filter hybridization. However, they give only a rough estimate of the T_m for filter hybridizations. As a consequence of binding nucleic acid to filters, the T_m of hybrids is often lower than would be predicted from solution hybridization studies [2].

In practice, for perfectly matched hybrids, DNA:DNA hybridizations are carried out at 60–70°C in aqueous solution and 37–45°C in the presence of formamide. Washing takes place in aqueous solution at 5–25°C below the T_m. For imperfectly matched hybrids, the temperatures of both hybridization and washing solutions are lowered to stabilize the hybrids. Thus, hybridization takes place at temperatures of 35–42°C in formamide-containing solutions: washing is in aqueous solutions at 20–30°C below the T_m.

DNA:RNA and RNA:RNA hybridizations are usually carried out in 50% formamide at an incubation temperature of 45–65°C. Because these hybrids have a higher T_m than DNA:DNA hybrids, washing conditions can be more stringent, for example 60–65°C in 0.1–1 × SSPE. High stringency washing is important for RNA:RNA hybridizations because some probes show spurious hybridization to ribosomal RNA that is not shown by DNA probes from the same region.

7.2.7 Formamide

Formamide decreases the T_m of DNA in solution by about 0.63°C for every 1% formamide [9]. The effect is less pronounced for DNA:RNA and

RNA:RNA hybrids and is not linear with formamide concentration. The depression of the T_m by formamide is a most useful property because by including 30–50% formamide in hybridization reactions, the T_m is lowered by 19–32°C which means that the hybridization temperature can be lowered by the same amount. This has several practical benefits. At lower temperatures nucleic acid is retained better on the filter, nitrocellulose filters are more robust and nucleic acid, particularly RNA, is less likely to undergo thermal degradation.

In the presence of 80% formamide, DNA:RNA hybrids are more stable than their DNA:DNA counterparts by some 10-30°C depending on the sequence. This property can be exploited in Northern blots: by hybridizing in the presence of 80% formamide preferential formation of DNA:RNA hybrids is promoted and reassociation of the DNA probe is prevented.

Concentrations of formamide between 30 and 50% do not appear to reduce the rate of filter hybridization, whereas 20% formamide reduces it by about one-third [9]. The reason for the anomalous depression of the rate at low formamide concentration is not known. The effect that high concentrations of formamide have on the rate of filter hybridizations is also not known, but solution hybridization studies show a threefold depression in rate for DNA:DNA hybridizations and 12-fold for RNA:RNA hybridizations in 80% formamide [10]. Qualitatively similar effects probably occur in filter hybridization.

In practice, hybridizations for perfectly matched hybrids can be carried out in either aqueous or formamide-containing solutions. There are no hard and fast rules for choosing which to use. Investigators often choose one on an arbitrary basis and if results are not satisfactory, they try the other. Hybridization of imperfectly matched sequences is carried out in formamide-containing solutions.

Formamide can be used to control the stringency of hybridization [9]. By holding the temperature constant and adding different concentrations of formamide, different effective temperatures can be attained (*Table 7.1*). For example, the T_m of SV40 virus DNA in aqueous solution is 93°C. In the presence of 30% formamide, the T_m drops to 75°C and if hybridization is carried out at 35°C this is equivalent to $(75-35)°C = 40°C$ below the T_m of perfectly matched hybrids. Temperatures that are effectively 46°C below the T_m of perfectly matched duplexes can be attained. This is very useful for detecting imperfectly matched (distantly related) sequences as it allows detection of sequences which differ by as much as 35% of their sequence. This is the upper limit to the degree of mismatching that can detected because above this value the probe will detect short perfectly matched sequences that occur by chance alone [11].

Table 7.1. Effective temperatures reached by varying the concentration of formamide and holding the T_i constant

% Formamide	T_m (°C)	T_i (°C)	Effective temperature if T_i = 35°C (°C)
0	93		
20	81	35	T_m −46
30	75	35	T_m −40
40	68	35	T_m −33
50	62	35	T_m −27

T_m is calculated from the relationship:

$$T_m = 81.5°C + 16.6 \log_{10} \left(\frac{[Na^+]}{1+0.7[Na^+]} \right) + 0.41 \ (\% \ G+C) \ -500/L$$
$$-P-0.63 \ (\% \ formamide)$$

where T_m is the melting temperature, $[Na^+]$ is the monovalent cation concentration, (% G + C) is the percentage of (guanine + cytosine residues) in the DNA, L is the length of the duplex in bp and P is the percentage mismatching.

In *Table 7.1* the T_m of SV40 DNA is calculated assuming: a cation concentration of 1 M; %(G + C) of 41%; L = 500; P = 0.

7.2.8 Imperfectly matched sequences

Data from solution hybridization studies show that under optimal conditions mismatching decreases the rate of hybridization about two-fold for every 10% mismatching [4]. The T_m of a hybrid is also depressed by mismatching because there are fewer hydrogen bonds to be disrupted than in a perfectly matched duplex. The reduction is estimated to be about 1°C for every 1–1.7% loss in homology [4,12]. However, the stability of hybrids also depends on the distribution of mismatched bases in the duplex. For example, a duplex with 20% mismatching will be very unstable if every fifth base is mismatched whereas a duplex also with 20% mismatching will be far more stable if the mismatches are clustered. Furthermore, a cluster of mismatches at one end of a molecule will reduce the T_m by less than a cluster of mismatches at the center.

Many applications of hybridization involve hybridizing to sequences that are related, but not identical, for example, members of multigene families or genes in other species. Hybridization conditions must be selected that allow imperfectly matched hybrids to form and washing conditions must be found that allow them to persist while reducing the background to an acceptable level (*Figure 7.1*). Such conditions have to be determined by trial and error, but initially hybridization is carried out at low stringency using different concentrations of formamide (e.g. 30–50%) at a fixed low temperature (e.g. 37°C) and washing at high salt concentration of salt and low temperatures (e.g. 6 × SSC, 40°C). These

(a) **(b)**

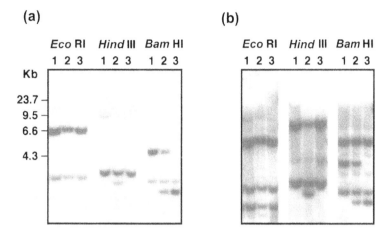

Figure 7.1. Detection of related sequences. Duplicate Southern blots containing 5 μg restricted human DNAs/lane were probed with an Ig V$_\lambda$ probe. (a) High stringency hybridization and washing which detects only well-matched hybrids. (b) Low stringency hybridization and washing which detects distantly related as well as closely related sequences. High stringency hybridization was in a 50% formamide, 5 × SSC, 5 × Denhardt's-containing solution for 16 h at 42°C. The final washes were in 0.1 × SSC, 0.1% SDS at 65°C. Low stringency hybridization was under the same conditions except that the solution contained 30% formamide. The final washes were in 2 × SSC, 0.1% SDS at 50°C.

variables are altered one at a time until a satisfactory hybridization signal is obtained.

7.2.9 Hybridization accelerators

The presence of inert polymers such as 10% dextran sulfate or 10% polyethylene glycol increases the rate of filter hybridization with both DNA and RNA probes [13,14]. For single-stranded probes the rate increases by up to four-fold whereas for double-stranded probes the increase is up to 12-fold. The polymer is thought to exclude the probe from the volume that it itself occupies and thus effectively increases the concentration of the probe.

Polymers increase not only the rate of reaction, but also the extent, i.e. more probe appears in hybrids. However, this effect is probably caused by the formation of networks of reassociated probe which, by virtue of single-stranded regions, hybridize to the target on the filter. This effect may be useful in amplifying the signal on the filter for qualitative hybridization, but should be avoided for quantitative analyses. Formation of networks can be minimized by using short probes that lack self-complementary regions and by using short incubation times because networks form late in the hybridization reaction [13].

There is no advantage in using hybridization accelerators unless the rate of hybridization is low, the target sequences on the filter are rare or the concentration of probe sequences is low. Solutions containing dextran sulphate or polyethylene glycol are viscous and filters have a tendency to stick to the hybridization vessel through surface tension. So nitrocellulose filters, in particular, need to be handled very carefully to prevent tearing. Higher backgrounds tend to be found when hybridization reactions contain these polymers, but this may merely reflect inadequate washing of the filters.

Hybridization accelerators such as phenol and guanidinium isothiocyanate that are useful for solution hybridizations [15] are not used in filter hybridization because they tend to bind probe nonspecifically to the filter causing high backgrounds.

7.2.10 *Enzyme-linked probes*

Enzyme-linked probes are useful because they shorten the time of detection (see Chapter 13). However, conditions of hybridization and washing must take into account the stability of enzymes linked to nucleic acids [16]. The T_m of oligonucleotides is unchanged by the attachment of hydrogen peroxidase (HRP), but is decreased by about 10°C if alkaline phosphatase is bound. To preserve enzyme activity, horseradish peroxidase labeled probes are incubated in high concentrations of urea at temperatures of 37–50°C. This may compromise specificity. Alkaline phosphatase is more stable than HRP and can be incubated at 50–60°C for 1–2 h without significant loss of enzyme activity.

7.3 Reaction volume

Reaction volumes are kept low – in general the lower the better – as this effectively concentrates the probe. This is especially important if the probe is scarce. However, it is important to ensure that the volume is sufficient to keep the filters covered with fluid during hybridization otherwise the probe will dry on to the filter and cause high background.

7.4 Time of hybridization

An estimation of how long to carry out hybridization can be made by considering how quickly the probe is used up. The calculations are different depending on whether or not the probe can reassociate.

If the probe is double-stranded, reassociation in solution will take place at the same time as hybridization to the filter-bound target and may even be faster. By the time that reassociation in solution has reached 1–3 × $C_0 t_{1/2}$ most of the probe (between 50 and 75%) will be double-stranded and the amount available to hybridize at the filter will be negligible (see Section 3.2.1). The $C_0 t_{1/2}$ for reassociation of the probe in solution can be derived from the following relationship [7]:

1 µg of denatured double-stranded DNA with a complexity of 5 kb and in 10 ml of hybridization solution reaches $C_0 t_{1/2}$ in 2 h.

To determine the number of hours (H) for any other probe to reach $C_0 t_{1/2}$, the relevant figures can be substituted in Equation 7.4:

$$H = \frac{1}{X} \times \frac{N}{5} \times \frac{Z}{10} \times \frac{2}{1} \tag{7.4}$$

where X is the weight of the probe added (in µg), N is the complexity of the probe (which for most probes can be taken as its length in kb) and Z is the volume of the reaction (in ml).

Additional factors can be included in the above equation to take account of different conditions, for example in 20% formamide, the reaction rate in solution is depressed three fold, so it will take three times as long to reach $C_0 t_{1/2}$ [9].

If the probe is in excess and cannot reanneal, the reaction follows pseudo-first order kinetics. The time in seconds for half the probe to hybridize can be estimated by the relationship:

$$t_{1/2} = \frac{N \ln 2}{3.5 \times 10^5 \times L^{0.5} \times C_0} \tag{7.5}$$

where C_0 is the probe concentration (mol. nucleotide/1^{-1}), L is the length of probe fragments (nt) and N is the complexity of the probe (nt).

This relationship predicts that hybridization will be speeded up by using a high concentration of probe of low complexity. The rate of a hybridization-limited filter hybridization is 10 times slower than that in solution [1] so for these conditions, the time derived in Equation 7.5 should be multiplied by 10 to estimate the $t_{1/2}$. The time of hybridization need not be prolonged longer than the time required to reach 3 × $C_0 t_{1/2}$. By this time 75% of the probe will have been used up and the reaction will essentially be complete.

In practice, DNA:DNA hybridizations are usually carried out for 6–18 h RNA:DNA hybridizations for 8–18 h and RNA:RNA hybridizations for 10–12 h [7]. For convenience, hybridizations are often carried out overnight although this is probably much longer than is necessary. It should be noted, that in certain circumstances overlong hybridization may be detrimental as when discriminating between closely and distantly related members of a gene family. Discrimination is usually better at short hybridization times [2].

References

1. **Flavell, R.A., Birfelder, E.J., Sanders, J.P. and Borst, P.** (1974) *Eur. J. Biochem.* **47:** 535–543.
2. **Beltz, G.A., Jacob, K.A., Eickbush, T.H., Cherbas, P.T. and Kafatos, F.C.** (1983) *Methods Enzymol.* **100:** 266–285.
3. **Flavell, R.A., Borst, P. and Birfelder, E.J.** (1974) *Eur. J. Biochem.* **47:** 545–548.
4. **Wetmur, J.G. and Davidson, N.** (1968) *J. Mol. Biol.* **31:** 349–370.
5. **Bonner, T.I., Brenner, D.J. Neufield, B.R. and Britten, R.J.** (1973) *J. Mol. Biol.* **81:** 123–135.
6. **Birnsteil, M.L., Sells, B.H. and Purdom, I.F.** (1972) *J. Mol. Biol.* **63:** 21–39.
7. **Sambrook, J., Fritsch, E.F. and Maniatis, T.** (1989) In: *Molecular Cloning: A Laboratory Manual, 2nd Edn.* Cold Spring Harbor Laboratory Press, Cold Spring Harbor, NY.
8. **Wetmur, J.G.** (1991) *Crit. Rev. Biochem. Mol. Biol.* **26:** 227–259.
9. **Howley, P.M., Israel, M.F., Law, M-F. and Martin, M.A.** (1979) *J. Biol. Chem.* **254:** 4876–4883.
10. **Casey, J. and Davidson, N.** (1977) *Nucleic Acids Res.* **4:** 1539–1552.
11. **Lathe, R.** (1985) *J. Mol. Biol.* **183:** 1–12.
12. **Caccone, A., DeSalle, R. and Powell, J.R.** (1988) *J. Mol. Evol.* **27:** 212–216.
13. **Wahl, G.M., Stern, M. and Stark, G.R.** (1979) *Proc. Natl Acad. Sci. USA* **76:** 3683–3687.
14. **Amasino, R.** (1986) *Anal. Biochem.* **152:** 304–307.
15. **Kohne, D.E., Levinson, S.A. and Byers, M.J.** (1977) *Biochemistry* **16:** 5329–5341.
16. **Kessler, C.** (1995) In: *Gene Probes 1: A Practical Approach* (B.D. Hames and S.J. Higgins). IRL Press, Oxford, pp. 93–144.

8 Hybridization strategy: oligonucleotide probes

8.1 Introduction

Hybrids containing oligonucleotides are much less stable than hybrids of long nucleic acids. This is reflected in lower melting temperatures. The instability of the hybrids is one of the most important factors to be considered when designing oligonucleotide hybridizations. Hybridization and washing conditions are significantly different from those used with long probes.

8.2 Factors affecting the hybridization of oligonucleotides and stability of hybrids

8.2.1 Melting temperature

The T_m of oligonucleotide hybrids is defined as the temperature at which 50% of the hybrids have dissociated. At T_m, there is equilibrium between the dissociation and association of the hybrid (*Figure 1.5b*).

The T_m is profoundly dependent on length, base composition and base sequence of the oligonucleotide. The latter effect arises because the stacking forces between neighboring bases depend on base sequence and are relatively more important in a very short molecule than in a long one where the sequence effect tends to even out.

For short oligonucleotides (11–23 nt) in 1 M salt and for perfectly matched hybrids, the T_m can be estimated using the relationship [1]:

$$T_m = 4(G+C) + 2 (A+T) \tag{8.1}$$

where T_m is the temperature (°C) at which 50% hybrids have dissociated and G, C, A and T are the number of guanine, cytosine, adenine and thymine bases, respectively, in the oligonucleotide.

The relationship above overestimates the T_m for longer nucleotides and is generally replaced by Equation 8.2 [2]:

$$T_m = 81.5°C + 16.6 (\log_{10}[M]) + 0.41 (\%G+C) - 600/L \qquad (8.2)$$

where M is the monovalent cation molar concentration, (%G+C) is the content of guanine and cytosine residues and L is the length in nucleotides of the oligonucleotide. This relationship holds for oligonucleotides of 14 to at least 72 nt. Note that although the above relationships are useful for giving an estimate of melting temperatures, the T_m found empirically is often higher by up to 10°C than that calculated. This probably arises through one of the strands of the hybrids being immobilized [3].

In practice, hybridization with oligonucleotide probes is usually carried out at 5–10°C below the T_m. These conditions are stringent and will allow perfectly matched hybrids to form, but will select against hybrids containing mismatched bases.

Effect of quaternary ammonium salts on T_m. The effect of base composition of oligonucleotides on T_m of hybrids can be suppressed by hybridizing in the presence of quaternary ammonium salts such as 2.4 M tetraethylammonium chloride (TEACl) or 3 M tetramethylammonium chloride (TMACl) [4,5]. These salts bind selectively to A:T base pairs preventing them melting at low temperature. So in these solvents, the T_m of hybrids is independent of base composition and is dependent on length only. This holds for oligonucleotides of up to about 200 nt, but the T_m is essentially independent above 200 nt in length (*Figure 8.1a*).

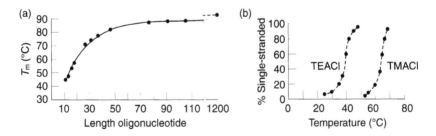

Figure 8.1. Effect of tetraalkylammonium salts on T_m of oligonucleotide hybrids. (a) T_m in 3M TMACl as a function of oligonucleotide length. (b) Denaturation of 19-mer oligonucleotide hybrids in the presence of 2.4 M TEACl and 3M TMACl.

In 2.4 M TEACl, the T_m of hybrids is about 10–20°C lower than in 3 M TMACl (*Figure 8.1b*). Hybridizing in the presence of TEACl is useful because the conditions are less harsh and filters are less likely to disintegrate. On the other hand, in 3 M TMACl the incubation temperatures are higher and there is less nonspecific binding of the probe to the filter, so the background is lower.

With a single oligonucleotide probe of known sequence and up to about 16 nt in length, there is no advantage in using quaternary ammonium salts instead of SSC-containing solutions. The main benefit of using quaternary ammonium salts is when several oligonucleotide probes are used at once. For example, pools of oligonucleotides are frequently used as probes to screen recombinant libraries. Since these pools will not have a single consensus T_m, it is difficult to decide what T_i to use. The problem of base composition can be overcome by hybridizing in the presence of quaternary alkylammonium salts since in this solvent only the length of oligonucleotide is important. The appropriate T_m can be deduced from *Figure 8.1* or published sources [4,5] and an incubation temperature of about 10°C below the calculated T_m can be used as a starting point for optimizing the hybridization signal.

8.2.2 Mismatches

Mismatches substantially reduce the thermal stability of oligonucleotide hybrids particularly for shorter oligonucleotides [6]. The effect can be estimated roughly from the relationship that every 1.7% mismatching reduces the T_m of a hybrid by 1°C [7]. However, other factors affect the reduction in stability.

- A mismatch at the center of a hybrid reduces the T_m by more than one at the end.
- The identity of the bases at the position of mismatch affects the stability of the hybrid. Some mismatched base pairs are fairly stable (e.g. G-T). The order of stability of oligonucleotide-DNA with different mismatched bases at the same position is (from most to least stable) T-A, A-T > G-T, G-A, > A-A, T-T, C-T, C-A [8].
- If there is more than one mismatch, hybrids are less stable if the mismatches are dispersed rather than clustered.

For oligonucleotides shorter than about 20 nucleotides, a single mismatched base pair can reduce the T_m of the hybrid by as much as 10°C [6]. The difference in thermal stability between an oligonucleotide: DNA hybrid with a mismatch and its perfectly matched counterpart can be exploited experimentally to detect mutations [9] (see Section 15.5.4). This application is used extensively in prenatal diagnosis of hereditary disease.

8.2.3 *Rate of hybridization*

Single oligonucleotide probes hybridize rapidly to their targets. There are several reasons. Most oligonucleotide probes are available in large amounts, so it is easy to achieve high concentrations of probe in solution. In addition, being small, oligonucleotides diffuse readily to the target. They do not (or should not) reassociate, so there is no competing reaction in solution to reduce the concentration of probe. Finally, the complexity is low.

However, with probes consisting of mixtures of oligonucleotides, the complexity is high and long periods of hybridization are required.

8.2.4 *Salt concentration*

Although the stability of oligonucleotide hybrids depends on the concentration of salt [10], salt concentration is not generally used to control stringency of hybridization. Hybridizations are usually carried out in either 6 × SSC or 6 × SSPE. The salt concentration of washing may be reduced to 2 × SSC or 2 × SSPE.

8.3 Time period for hybridization

Because hybrids containing oligonucleotides are unstable even when they are perfectly matched, hybridization times are usually kept to a minimum. This is not a problem when using single oligonucleotide probes because the rate of hybridization is rapid (Section 8.2.3). So with a high concentration of oligonucleotide probe, short hybridization times can be used to detect even single-copy target sequences.

Oligonucleotide hybridization follows pseudo-first order kinetics so the time required for half completion of the reaction can be calculated from Equation 8.3 (which is the same as Equation 7.5):

$$t_{1/2} = \frac{N \ln 2}{3.5 \times 10^5 \times L^{0.5} \times C_o} \tag{8.3}$$

where $t_{1/2}$ is the time in seconds required for the reaction to be 50% complete, N is the complexity of the probe, L is the length of the probe in nucleotides and C_o is the concentration of the probe (mol. nucleotide 1^{-1}). This equation is valid for oligonucleotides as short as 11 nt. Note that the concentration of oligonucleotides is usually expressed in terms of the strand, rather than number of nucleotides. To convert into the concentration of nucleotides, multiply the concentration by the oligonucleotide length, for example 2 µM of a 20-mer = 40 µM nucleotides.

For a probe of 17 nucleotides (complexity) present at $20\,\mathrm{ng\,ml}^{-1}$ (6×10^{-8} mol nucleotide l^{-1}), the hybridization should be 50% complete in $136\,\mathrm{s}$. However, the actual rate of hybridization on filters is about 3–4 times slower than the calculated rate so in the above example, the reaction will be 50% complete in about $9\,\mathrm{min}$. The reaction will have reached equilibrium by 2–3 h, so there is no advantage in hybridizing longer than this.

With mixtures of oligonucleotides, the complexity of the pool is high and at an overall probe concentration of $20\,\mathrm{ng\ ml}^{-1}$, each oligonucleotide will be present at a much lower concentration than in the above example. So incubation has to be prolonged – perhaps for 1–2 days [11].

8.4 Hybridization accelerators

Accelerators such as polyethylene glycol or dextran sulfate are not used for oligonucleotide hybridizations. The rate of hybridization is not limiting for single sequence oligonucleotides and in any event the main effect of accelerators is in promoting network formation which will not occur with such short probes.

8.5 Washing

Hybrids containing oligonucleotides may be unstable even at $T_{\mathrm{m}} - 10°\mathrm{C}$. If an oligonucleotide dissociates from its complement during washing, dissociation is essentially irreversible. This is because the concentration of free oligonucleotide is too low for rehybridization and the concentration is reduced even further each time the washing solution is replaced. So, to minimize dissociation of hybrids, washing times are kept short – sometimes only to a few minutes.

The temperature of washing is important in determining which hybrids persist. Lathe [12] has derived an empirical equation for estimating T_{w}, the recommended final stringent washing temperature, in degrees centigrade in $2 \times \mathrm{SST}$:

$$T_{\mathrm{w}} = 94 - \frac{820}{L} - 1.2(100 - h) \tag{8.4}$$

Where the value of $94°\mathrm{C}$ is the calculated T_{m} of long random DNA with a 50% (G+C) content, L is the length of the probe and h is the % homology between the probe and its target.

As $2 \times SST$ has an ionic strength close to $2 \times SSC$. The washing temperature in $2 \times SST$ can be converted into the equivalent in $6 \times SST$ by adding 8°C [12].

Figure 8.2 shows the calculated T_w for different lengths of oligonucleotide. Note how mismatching reduces the T_w. The calculated T_w for a 30-mer oligonucleotide with 90% homology is the same as that of a perfectly-matched 21-mer.

8.6 Control of stringency

In practice the specificity is usually controlled by manipulating the temperature of hybridization and the temperature, salt concentration and time of washing (*Table 8.1*).

8.7 Optimizing conditions

It is most important for oligonucleotide hybridizations that pilot experiments are first carried out to check and optimize the selected conditions.

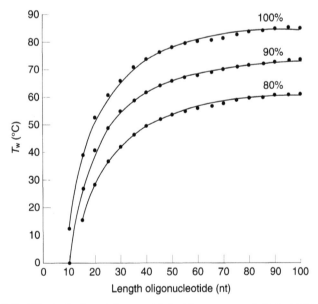

Figure 8.2. Effect of mismatches on high stringency wash temperature for targets sharing 100%, 90% or 80% homology with an oligonucleotide probe.

Table 8.1. Controlling the stringency of oligonucleotide hybridizations

	Hybridization	Washing
High stringency	Incubation temperature close to T_m High salt concentration (6 × SSC)	Incubation temperature close to T_m Low salt concentration (2 × SSC)
Low stringency	Incubation temperature 10°C or more below T_m High salt concentration (6 × SSC)	Incubation temperature 10°C or more below T_m High salt concentration (6 x SSC)

Slight changes in hybridization and washing conditions can have a major effect on which hybrids form and which are allowed to persist.

An initial set of conditions is chosen that is estimated to be appropriate for the proposed experiment. Hybridization is carried out and the result monitored. A further series of hybridizations is carried in which one variable is altered, for example a range of hybridization temperatures may be tested. The temperature which gives the best hybridization signal is now kept constant where another variable such as salt concentration of washing or temperature of washing is tested. The process of optimizing one variable at a time is continued until the hybridization signal is no longer improved.

8.8 Probe design

The design of the probe is one of the most important factors in determining the success or failure of oligonucleotide hybridizations.

8.8.1 Length of the probe

To be useful as a probe, an oligonucleotide must be sufficiently long to hybridize specifically to the target sequence, but not to unrelated sequences. If it is too short, the probe will hybridize to many sites in the target and there will be little or no specificity. The minimum size required for a probe to be unique depends on the complexity of the sequences being probed. It can be determined from the relationship:

$$F = (1/4)^L \times 2N \qquad (8.5)$$

where F can be regarded as either, the frequency with which a sequence complementary to the probe exists by chance in the bank of sequences being screened, or the expected number of perfect matches in the bank being screened. L is the oligonucleotide length and N is the complexity of the target [13]. (The letters in the above relationship have been changed from the original in order to maintain consistency of meaning throughout this book.)

The reason that the term $2N$ is used rather than N alone is that the sequence in each strand of double-stranded DNA is different and the oligonucleotide could potentially hybridize to either. As the value of L increases, the number of expected matches decreases exponentially. For an oligonucleotide to be useful, the value of F should not be greater than 1 and preferably less than 0.1.

A different form in which this relationship is sometimes expressed is when $4^L = 2N$, a sequence of length, L, would be expected to occur only once in a genome. So for a sequence to be unique, 4^L must be greater than $2N$.

To determine the minimal length of oligonucleotide required for a given experiment, Equation 8.5 is solved for F using the known complexity of the nucleic acid sample and different values of L. The lowest value of L for which F is less than 1 gives the minimum length for an oligonucleotide to be present once in the bank of sequences being screened.

As examples:
- The complexity of the prokaryotic genome is almost the same as its size (Section 3.3). For *E. coli* the genome size is 4.2×10^6 bp and $2N = 8.4 \times 10^6$ nt. So, in order to be present only once, the expected minimum length of oligonucleotide is 12 nt (i.e. at values of L below 12, F is more than 1 in Equation 8.5).
- The complexity of the mammalian genome approximates to 3×10^9 bp. Applying the above formula shows that an oligonucleotide sequence would have to be at least 17 nt long in order to be unique.
- The complexity of mRNA and cDNA derived from it is much less than that of genomic DNA because only a small proportion of the genomic DNA is transcribed. So the minimal size of oligonucleotide expected to be unique in screening a cDNA library will be smaller than that for screening the corresponding genomic DNA library.

Equation 8.5 is valid provided that there is a random distribution of bases in the target and the (G+C) content is 50%. For many genomes these provisos do not hold and a more general expression is used:

$$F = \{(g/2)^{(G+C)/L} \times [(1-g)/2]^{(A+T)/L}\}^L \times 2N \qquad (8.6)$$

Where g is the (G+C) content and A,G,C and T are the numbers of each nucleotide present in the probe [13]. If (G+C) = 50%, Equation 8.6 simplifies to:

$$F = (1/4)^L \times 2N$$

which is the same as Equation 8.5.

8.8.2 Single sequence probe

If the sequence of the DNA or RNA target is known precisely, then an exact complementary probe can be synthesized. Before actually synthesizing the oligonucleotide, however, it is prudent to carry out some checks:

- Run a computer search to check if the sequence is likely to be unique in the target. References to computer programs are given in Appendix C.
- Make sure that the proposed sequence is free of regions of self-complementarity which may inhibit both labeling and hybridization of the oligonucleotide.
- Avoid oligonucleotides that are rich in G as they are difficult to purify and handle.
- For detecting sequences on Northern or RNA dot blots, check that the probe has the sequence of the antisense strand.
- If mismatches (mutations) are to be detected in the target, two different probes are usually synthesized (see Section 15.5). One has the normal sequence and the other has the mismatched sequence. The position of the mismatched base in the oligonucleotide is important because this affects the thermal stability of imperfectly matched hybrids. The mismatched base should be in the middle to cause greatest reduction in the T_m. The identity of the mismatched base is also important as this affects the stability of the hybrid. Bases which will form G:T and A:G mismatches on hybridization should be avoided because these mismatched base pairs are fairly stable. If necessary prepare a probe that is complementary to the other strand so that a less stable C:A or T:C mismatch is generated instead. Note however, that a sense strand probe will not work for detecting mutations in mRNA. This problem can be avoided by probing cDNA instead of mRNA.

With a single sequence probe, there will be a single T_m which can be deduced from the relationships in Section 8.2.1. The rate of hybridization will be high because the complexity of the probe will be low.

8.8.3 Mixtures of oligonucleotide probes

If the probe sequence is deduced from protein sequence data, there will be several possible oligonucleotide sequences because of the degeneracy of the code. If all possible oligonucleotides are synthesized, the pool will have a very high complexity and the rate of hybridization will be very low.

The length of all the oligonucleotides in the pool will be the same, but the base composition and thus the T_m will be different for different members. It is unlikely that it will be known which oligonucleotide is the exact complement of the target sequence and since the pool of oligonucleotides will not have a unique T_m, there is a challenge to find optimal conditions for hybridization and washing of filters. There are two main ways of selecting hybridization conditions.

The first is to find conditions that are appropriate for the member with the lowest T_m which is generally the most AT-rich oligomer [11]. This procedure will detect not only perfectly matched hybrids, but also a large number of false positives. These arise because conditions which are appropriate for perfectly matched hybrids with a high (A+T) content may also allow formation of hybrids with a higher (G+C) content, but which contain a mismatch. If the number of positively scoring hybrids is low, false positives may be detected by direct sequencing or by hybridizing with a second oligonucleotide, preferably from a different part of the protein.

An alternative and easier procedure is to take advantage of the fact that in the presence of tetraalkylammonium salts the T_m of hybrids depends on the length of the oligonucleotide, but not on its composition. Hybridization can be carried out in the presence of $6 \times$ SSC with washing in tetraalkylammonium salts at 5°C lower than the T_m. Alternatively, more stringent conditions can be used by both hybridizing and washing in the presence of tetraalkylammonium salts at 5°C lower than the T_m.

This procedure will also give rise to false positives since perfectly matched hybrids formed with any of the oligonucleotides in the pool will be stable.

8.8.4 Reducing the complexity of an oligonucleotide pool

If using a mixture of oligonucleotides to probe a target, it is best to design the mixture from that part of the protein sequence for which there is least codon redundancy. This will reduce the number of oligonucleotides necessary to cover all possible sequences. The number of oligonucleotides can be further reduced by incorporating inosine at positions of high redundancy because inosine can pair with G, C and A leading to a more stable hybrid than when a mismatched base pair is present [14].

Whenever possible, the complexity of oligonucleotide pools should be kept as low as possible because the rate of hybridization decreases as the complexity of the probe increases. So, if it is necessary to use many oligonucleotides for screening, it is preferable to construct several pools of up to 20 oligonucleotides each than one pool containing all possible oligonucleotides.

An alternative means of reducing the complexity of an oligonucleotide pool is to design a single, longer oligonucleotide with an 'optimized' sequence. This takes into account features such as the known codon preferences of the species, the under-representation of the dinucleotide CpG in many genomes and the avoidance of coding sequences that involve leucine, arginine and serine, all of which have six codons. A detailed consideration of the theoretical and practical aspects of this approach and of methods to reduce the number of false positives are given in reference [12].

References

1. **Suggs, S.V., Hirose, T., Miyake, T., Kawashima, E.H., Johnson, M.J., Itakura, K. and Wallace, R.B.** (1981) *ICN-ICLA Symp. Mol. Cell. Biol.* **23:** 682–693.
2. **Bolton, E.T. and McCarthy, B.J.** (1962) *Proc. Natl Acad. Sci. USA* **48:** 1390–1397.
3. **Albertsen, C., Haukanes, B.-I., Aasland, R. and Kleppe, K.** (1988) *Anal. Biochem.* **170:** 193–202.
4. **Wood, W.I., Gitschier, J., Lasky, L.A. and Lawn, R.M.** (1985) *Proc. Natl Acad. Sci. USA* **82:** 1585–1588.
5. **Jacobs, K.A., Rudersdorf, R., Neill, S.D., Dougherty, J.P., Brown, E.L. and Fritsch, E.F.** (1988) *Nucleic Acids Res.* **16:** 4637–4650.
6. **Wallace, R.B., Shaer, J., Murphy, R.F., Bonner, J., Hirose, T. and Itakura, K.** (1979) *Nucleic Acids Res.* **6:** 3543–3557.
7. **Caccone, A., DeSalle, R. and Powell, J.R.** (1988) *J. Mol. Evol.* **27:** 212–216.
8. **Ikuta, S., Takagi, K., Wallace, R.B. and Itakura, K.** (1987) *Nucleic Acids Res.* **15:** 797–811.
9. **Wallace, R.B., Johnson, M.J., Hirose, T., Mikaye, T., Kawashima, H. and Itakura, K.** (1981) *Nucleic Acids Res.,* **9:** 879–894 .
10. **Wetmur, J.G. and Davidson, N.** (1968) *J. Mol. Biol.* **31:** 349–370.
11. **Sambrook, J., Fritsch, E.F. and Maniatis, T.** (1989) *Molecular Cloning: a Laboratory Manual.* Cold Spring Harbor Laboratory Press, Cold Spring Harbor, New York.
12. **Lathe, R.** (1985) *J. Mol. Biol.* **183:** 1–12.
13. **Nei, M. and Li, W.H.** (1979) *Proc. Natl Acad. Sci. USA* **76:** 5269–5273.
14. **Ohtusa, E. Matsuki, S., Ikehara, M., Takahashi, Y. and Matsubara, K.** (1985) *J. Biol. Chem.* **260:** 2605–2608.

9 Choice of probe

9.1 Introduction

In order to interpret a hybridization experiment, hybrids must be distinguishable from nucleic acids that remain single-stranded. This is accomplished by labeling one of the reacting nucleic acids, the probe, with a reporter molecule. After hybridization, the probe that has not reacted is washed away and the hybrids are recognized by virtue of the reporter molecule.

In principle, any nucleic acid – double-stranded DNA, single-stranded DNA, oligonucleotides, mRNA and RNA – can act as a probe. The nucleic acid can be a single sequence or a mixture. The choice is determined by the purpose of the experiment and by the materials available. It is important when choosing a probe to check that the level of detectability desired, the type of label, the hybridization and washing conditions to be used and the proposed detection system are all compatible.

9.2 Characteristics of probes

9.2.1 DNA probes

Double-stranded DNA probes are very commonly used and consist of double-stranded DNAs that have been denatured and added quickly to the hybridization mix before they have time to reassociate. They are often cloned sequences and are of low sequence complexity.

Single-stranded DNA probes are usually derived by reverse transcription of RNA. If the template is mRNA, the cDNA will consist of many different sequences and the complexity of the DNA will be high. The individual species in the cDNA mix will be present at different concentrations and if the sequence of interest is very rare in the probe, it may be difficult or impossible to detect the target by filter hybridization.

Single-stranded probes may also be derived from cloned sequences in single-stranded phages such as M13. However, this source of DNA probes is less common than it used to be.

9.2.2 RNA probes

Single-stranded probes may be prepared by labeling isolated RNA, but it is more common to transcribe cloned sequences from recombinant plasmids in which the cloned sequences are flanked by promoters for bacteriophage RNA polymerases (see Section 11.3.3). Each enzyme is highly specific in that it will transcribe only from its own phage promoter and not a bacterial one. So, it is easy to produce large amounts of pure RNA by run-off transcription. The RNA is generally of low complexity and because it is single-stranded, there is no competing reassociation in solution during hybridization.

The advantages of RNA probes are that they are relatively easy to prepare and RNA:DNA and RNA:RNA hybrids are more stable than the corresponding DNA:DNA hybrids. However, their main disadvantage is that they are more difficult to handle than DNA probes because of the widespread presence of ribonucleases.

9.2.3 Oligonucleotide probes

Oligonucleotide probes are short (typically 17–50 nt in length) and are usually chemically synthesized. In theory they may be oligodeoxyribo-nucleotides, oligoribonucleotides or molecules with a modified backbone such as peptide nucleic acids. However, the vast majority of the published applications have used oligodeoxyribonucleotide probes and this is the only type of oligonucleotide discussed in detail here. (But see Chapter 16.)

Oligonucleotides may consist of a single specific sequence or, where the sequence is derived from protein sequence data, they may be composed of a mixture of sequences (see Section 8.8). They are easy to prepare in large quantities either in-house using a synthesizer or custom-made commercially. They are easy to label.

One of the most useful properties of oligonucleotide probes is that they can discriminate between target sequences that differ by as little as a single nucleotide. The main disadvantage in their use is that hybrids containing oligonucleotides are much less stable than hybrids containing longer nucleic acids. So great care must be taken to optimize hybridization and washing conditions.

9.2.4 Characterizing the probe

It is important to characterize the nucleic acid used for the probe. If any repetitive elements are present (e.g. *Alu*I sequences) they should be removed, especially if the probe is to be used to detect sequences that are present in low copy number. If a probe containing both single-copy and *Alu*I sequences is used to probe a blot containing genomic DNA, the *Alu*I sequences will hybridize and by virtue of their numbers, will completely swamp the signal from the single-copy sequence. *Alu*I sequences are present at a frequency of about once in every 4000 nucleotides in the human genome, so many cloned sequences contain a copy.

If the probe is a cloned sequence, it is usually desirable to excise and purify it away from the vector. There are two main reasons for this. First, many bacteriophage DNAs and plasmids share certain sequences. So, if a phage or plasmid library is screened using a probe that contains the vector sequence, every plaque/colony will give a positive signal – not because each contains the desired sequence, but because the vector sequences are hybridizing to each other. Second, when a filter is to be re-probed, the first probe is usually removed. If removal is incomplete, then when the second probe is added, sandwich hybridization may occur. This means that vector sequences of the second probe hybridize to vector-containing tails of the remaining first probe leading to false positives. The problem is avoided if both probes lack vector sequences.

9.3 Labeling the probe

In principle, DNA and RNA probes may be labeled *in vivo* by adding a limiting amount of labeled precursor to growing cells. In practice, however, most probes are prepared *in vitro* which is usually faster and gives better control of specific activity. There are two main types of label, radioactive and nonradioactive. The requirements of the experiment will determine the type of probe and choice of label.

9.3.1 Radioactive probes: general characteristics

Traditionally probes for filter hybridizations have been labeled with radioactive isotopes and this procedure is still widely used. The most commonly used isotope for filter hybridization is ^{32}P as it is sensitive and gives reasonably good resolution. This means that it can be used to detect very small amounts of nucleic acid and it can resolve hybrids that are close to each other on the filter. ^{32}P is generally introduced enzymatically to probes via commercially available radioactive nucleotides: NTPs, dNTPs and ddNTPs [1,2]. Radioactive probes are easy to prepare and can give a wide range of specific radioactivities (radioactivity incorporated per unit mass – mg, μg etc.).

Detection of hybrids is both easy and sensitive; single-copy sequences can be detected in 1 μg DNA. Low backgrounds are generally achieved. Radioactive probes can be used with all types of filters and can be removed fairly efficiently to allow filters to be re-probed.

There are several disadvantages to radioactive probes. Precautions have to be taken to prevent radiation being a health hazard. ^{32}P has a short half-life (14.3 days) which means that probes must be prepared frequently. They are expensive to prepare when taking into account the costs of: the isotope itself; photographic film for detection; licenses to use radioactivity; disposal of waste material; hand-held mini monitors; badges to monitor emission of radioactivity and safety equipment such as perspex/lucite shields.

9.3.2 Nonradioactive probes: general characteristics

A variety of nonisotopic alternatives are available for labeling nucleic acids. When first introduced, these lacked the sensitivity of radioactive probes, but advances in detection methods have led to the development of nonradioactive probe/detection systems which rival the sensitivity of radioactive ones. Probes for filter hybridization are most commonly labeled with biotin, digoxygenin or fluorescein. The enzymes alkaline phosphatase and horseradish peroxidase are fairly common reporter molecules [3]. Dinitrophenol, bromodeoxyuridine and acridines are still sometimes used as labels, but mercury and 2-acetylaminofluorene have largely been superseded because of their toxicity.

Biotin is a small, water-soluble vitamin, digoxygenin is a steroid obtained from the plant genus *Digitalis* and fluorescein is a fluorescent dye. All three reporter molecules are commercially available coupled via a linker arm to uracil nucleotides, usually UTP, dUTP and ddUTP (*Figure 9.1*). The length of the spacer arm between the reporter molecule and the base is important. Enzymatic labeling of probes is usually most efficient if the spacer arm is very short (about four carbon atoms in length), but detection of hybrids is very inefficient with such short spacers because of steric hindrance (see Chapter 13). Nucleotide derivatives with spacer arms of 11, 16 or 21 carbon atoms are most often used to label probes.

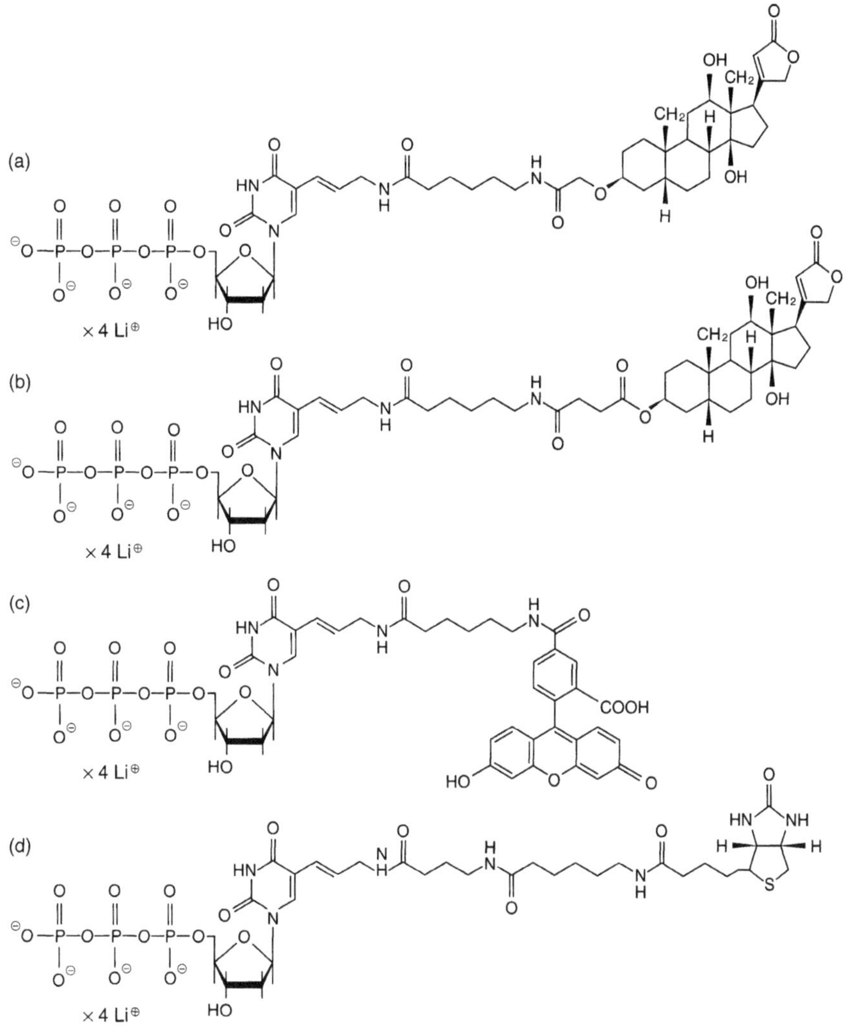

Figure 9.1. Structure of nonradioactively labeled nucleotides. (a) Alkali labile digoxygenin 11-dUTP. (b) Alkali stable digoxygenin 11-dUTP. (c) Fluorescein-12-dUTP. (d) Biotin 16-dUTP (re-drawn from Gene Probes 1: A Practical Approach, eds Hames and Higgins, 1995, by permission of Oxford University Press).

The same enzymatic methods that are used for radioactive probes can be used to introduce nonradioactive reporter molecules into RNA and DNA and at the 5′ or 3′ end or at internal positions [1,2]. Biotin and digoxygenin can also be added to nucleic acid by photochemical reaction to generate uniformly labeled probes [2,4]. Probes can also be labeled by PCR. There are two main approaches. The first uses prelabeled primers that have been synthesized using the appropriately modified phosphor-amidites. The second uses labeled dUTP derivatives to incorporate label during amplification. The main advantage of synthesizing probes by PCR is that large amounts of probe can be generated in a few hours from very small amounts of starting material which is particularly useful if the material is scarce.

Oligonucleotides can be labeled at any position during synthesis via the appropriate phosphoramidite derivative. If unlabeled oligonucleotides are available, they can be labeled by the same methods used to introduce ^{32}P into oligonucleotides except at the 5' end. Unlabeled oligonucleotides have to be activated at the 5' end before a reporter group can be attached. This usually involves adding a molecule with a free amino group. The activated oligonucleotide is reacted with a *N*-hydroxysuccinimidyl (NHS)-activated ester of the reporter molecule which covalently attaches the reporter to the amino group. The chemicals used in this process are unpleasant and toxic, but exposure can be limited by using preoptimized kits which are available commercially.

Enzyme reporter molecules are covalently attached to nucleic acids by chemical reactions such that they retain enzymatic activity [3].

When designing a hybridization experiment, it must be remembered that the identity of the reporter group may necessitate the modification of the reaction conditions. For example, if the reporter molecule is attached to a nucleotide by an alkali-labile bond, the probe must not be denatured by alkali treatment. As another example, if an enzyme is attached to the probe, the temperature of incubation and washing must not be too extreme or enzyme activity, which is the basis of the detection of hybrids, will be lost. Factors affecting the choice of label are summarized in *Table 9.1*.

9.4 Choosing a probe

9.4.1 Sensitivity required

A key consideration in selecting a probe is the level of detectability (sensitivity) required. This will influence the choice of nucleic acid (DNA, RNA or oligonucleotide), the reporter molecule (radioactive, nonradioactive) and the means of labeling (end-labeling or uniform labeling).

Under optimized conditions, a uniformly labeled radioactive probe of high specific radioactivity ($> 1 \times 10^9$ cpm μg^{-1}) will detect a single-copy sequence in 1 μg of Southern-blotted DNA. Increasing the amount of

Table 9.1. Choice of label

Radioactive probes
Advantages
  High specific activities can be achieved and can be calculated accurately.
  Sensitivity is good. Amounts down to 10 fg DNA can be detected on filters.
  Low background.
  Probes can be removed and filters reprobed.

Disadvantages
  Short half-life means probes must be made regularly.
  Radioactive decay causes degradation of probe.
  Radiation emission is a health hazard.
  Cost.

Nonradioactive probes
Advantages
  In general the health hazards are less than those associated with radioactive probes.
  Probes are stable and do not suffer from short half-lives.
  Probes can be made in bulk and stored. This allows uniformity of probe characteristics
    for use over a long period of time.
  In general detection is rapid.
  More than one probe can be hybridized at a time provided they are labeled with
    different labels, e.g. biotin and digoxygenin.

Disadvantages
  Many researchers feel that the sensitivity of detection is not as high as with
    radioactive probes.
  It is not as easy to determine the efficiency with which probes have been labeled
    as when using radioactivity.
  The presence of the reporter molecule may restrict the conditions used for hybridization.
  Backgrounds may not be as low as with radioisotopes, particularly if detection of hybrids
    is by dye precipitation.
  If dye-precipitation techniques are used to visualize hybrids, it is sometimes difficult to
    remove the probe or dye efficiently in order for the filter to be reprobed.

bound DNA to 10 μg or more will improve the chances of detection, but this may not be possible if the source of DNA is scarce.

With end-labeled probes, the amount of label per molecule of probe is much less than with uniformly-labeled probes. So to obtain a detectable signal, it may be necessary to increase the amount of nucleic acid bound to the filter. With oligonucleotide probes, their small size coupled to the limited amount of label that can be attached, may also necessitate increasing the amount of target nucleic acid bound to the filter.

The sensitivity of detection is influenced by the method used to detect hybrids. For example, nonradioactive probes may be detected by deposition of a colored precipitate or by chemiluminescence/ enhanced chemiluminescence (see Chapter 13). Light-generating detection methods are 10–100-fold more sensitive.

The amount of material bound to the filter will also influence the choice of probe. The more material that is bound to the filter, the easier it is to detect rare sequences. If the amount of nucleic acid available for binding is limited, then radioactive probes or probes that can be detected by enhanced chemiluminescence are probably the best choices.

9.4.2 DNA or RNA or oligonucleotide?

DNA probes are widely used as they are usually easy to prepare, easy to handle and give good levels of detection.

RNA probes can be used for the same applications as DNA probes. If screening Northern or RNA dot blots, care needs to be taken that the probe represents the antisense strand. DNA:RNA and RNA:RNA hybrids are more stable than DNA:DNA hybrids so hybridization and washing conditions can be more stringent. It can be difficult to remove RNA probes from filters because of the stability of RNA-containing hybrids.

Oligonucleotide probes are widely used for screening recombinant libraries, or hybridizing to dot and slot blots. Because they can discriminate between sequences that differ in a single nucleotide, they are particularly well-suited to detecting mutations in nucleic acid. Oligonucleotide probes are also very useful for probing nucleic acid in dried gels because their small size makes it easy for them to enter the gel.

9.4.3 End-labeling or uniform labeling?

Uniformly labeled probes are very widely used for screening libraries, probing Northern and Southern blots and for analyzing dot blots. More reporter molecules are available per probe molecule than when the probe is end-labeled, so less target material is needed to generate a detectable signal.

End-labeling is most useful with oligonucleotide probes or when mapping the ends of transcripts.

References

1. **Sambrook, J., Fritsch, E.F. and Maniatis, T.** (1989) *Molecular Cloning: a Laboratory Manual.* Cold Spring Harbor Laboratory Press, Cold Spring Harbor, New York.
2. **Hames, B.D. and Higgins, S.J.** (1995) *Gene Probes 1: A Practical Approach.* IRL Press, Oxford.
3. **Renz, M. and Kurz, C.** (1984) *Nucleic Acids Res.* **12:** 3435–3444.
4. **Forster, A.C., McInnes J.L., Skingle, D.C. and Symons, R.H.** (1985) *Nucleic Acids Res.* **13:** 745–761.

10 Basic techniques: binding of nucleic acid to filters

10.1 Types and properties of support

Several types of membrane are available to immobilize nucleic acids: the matrix of choice depends on the purpose of the experiment. Nitrocellulose, nylon and charged nylon are the most commonly used. The properties of these filters are summarized in *Table 10.1*.

Nitrocellulose was the first type of membrane to be used for filter hybridization and is still the matrix of choice for many applications. Its

Table 10.1. Properties of filters

Property	Nitrocellulose	Nylon	Charged nylon
Type of nucleic acid bound	DNA and RNA	DNA and RNA	DNA and RNA
Binding capacity	80–120 µg cm^{-2}	~450–600 µg cm^{-2}	~450-600 µg cm^{-2}
Pore sizes	0.45 µm for DNA 0.22 µm for < 500 nt DNA 0.1–0.22 µm for RNAs	0.45 µm 0.22 µm for short nucleic acid	0.45 µm
Ionic conditions used for transfer	High salt concentration	Wide range salt concentrations	Wide range salt concentrations
Binding procedure	Baking 80°C Noncovalent binding	Baking 80°C Noncovalent binding UV treatment Covalent binding Alkali treatment[a] Covalent binding	Baking 80°C Noncovalent binding UV treatment Covalent binding Alkali treatment Covalent binding
Advantages	Low backgrounds	High binding capacity Covalent binding	High binding capacity Covalent binding
Disadvantages	Fragility of unsupported nitrocellulose Noncovalent binding of nucleic acid		Higher backgrounds

[a]Some manufacturers do not recommend alkaline transfer with their brand of filter.

main advantage is that it gives low backgrounds, (i.e. nonspecific binding of probes to the filter is minimal) so that the hybrids can be more easily detected. It binds nucleic acid fairly efficiently except for small molecules of less than about 500 nt which bind rather poorly. Unsupported nitrocellulose is fragile, but supported versions are much more pliable and have the resolution and sensitivity of nitrocellulose alone. For repeated reprobings, supported nitrocellulose is preferred as the unsupported form tends to become brittle and to fragment.

Nylon and positively charged nylon filters have high tensile strength and are much more robust than unsupported nitrocellulose. They have high capacity for binding both DNA and RNA and are the preferred matrices if the target sequence is rare. In addition, they bind small nucleic acid molecules efficiently. Background levels with positively charged nylon filters tend to be higher than with nitrocellulose.

10.2 Transfer of nucleic acid to filters

All types of filters require the nucleic acid to be denatured for firm binding. No one method is best and the method of choice depends on the application [1]. The methods most commonly used are discussed below and protocols are given in Section 10.5.

10.2.1 Phage/colony transfer

Recombinant bacteriophage and colonies are grown on a plate under standard conditions. A sterile nitrocellulose or nylon filter is carefully laid on top and left for about 1 min to allow the phage/bacteria to stick to the filter (*Figure 6.1*). The filter is peeled off and inverted on successive pads that have been soaked in denaturing and neutralizing solutions. Alkali treatment is carried out both to lyse the phage/bacteria and to denature the DNA. Neutralization prevents alkali reacting with alkali-sensitive filters and is carried out for such a short time that reassociation of DNA is insignificant.

10.2.2 Transfer of DNA from gels

Capillary transfer. For DNA that has been fractionated by agarose gel electrophoresis, fragments are most commonly transferred to the filter by capillary blotting. This procedure is also known as Southern blotting after the inventor of the technique [2]. It is easy to set up and does not require elaborate or expensive equipment.

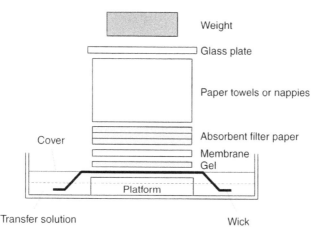

Figure 10.1. Capillary transfer of DNA from gel to filter. The gel is treated with alkali then neutralized. It is placed on a platform covered in absorbent paper that dips into transfer solution. A damp filter is placed on the gel, then successive layers of absorbent paper, paper towels, a glass plate and a weight. As transfer solution is drawn through the gel by capillary action, it carries DNA with it. On reaching the filter, the DNA binds.

The gel is treated with alkali to denature the DNA, then with neutralizing solution. A capillary transfer system is assembled as shown in *Figure 10.1*. The absorbent material above the filter draws the transfer solution in the reservoir up through the gel carrying the DNA with it. When the DNA reaches the filter it binds. The composition and ionic strength of the transfer solution are important for binding and are determined by the type and properties of the filter. Transfer is usually carried out overnight, but if the depth of the gel is sufficiently small, transfer may be complete in a few hours.

Long DNA (> 8 kb) does not migrate efficiently through agarose, so its size is first reduced by depurination under carefully controlled conditions. The gel is exposed to dilute acid at room temperature such that about 1 in every 500 purines is removed (see *Figure 10.2*). Since sites of depurination are sensitive to hydrolysis by alkali, exposure of depurinated DNA to alkali causes the DNA to fragment at these sites. The gel is then neutralized and capillary transfer carried out as above. Depurination does not affect the position of DNA in the gel, but simply improves the efficiency of transfer to the filter.

Vacuum blotting. The gel is placed on a filter which in turn is placed on a porous sheet in a special vacuum blotting apparatus. A gentle vacuum is applied which sucks transfer solution through the gel on to

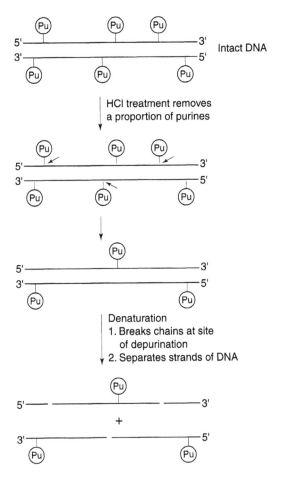

Figure 10.2. The size of DNA can be reduced by treatment with dilute acid followed by treatment with alkali. The size of DNA fragments depends on the concentration of acid and the length of time for which acid treatment is carried out.

the filter [3]. The process is fast and can be complete in 1–2 h. It gives good resolution and can be applied to both agarose and acrylamide gels, but it can be difficult to maintain an even pressure and there is a tendency for the gel to collapse if the suction is too strong. High and uneven backgrounds have been observed with RNA blots.

Positive pressure. Transfer can be achieved by applying positive pressure to the gel rather than vacuum [4]. Commercial equipment is available. Transfer buffers are driven through the gel carrying DNA with them on to the filter. The problem of gel collapse is reduced so higher pressures can be used which helps to increase the efficiency of transfer. The process takes less than an hour.

Electrophoretic transfer. Electrophoretic transfer is used mainly to transfer small DNA molecules from acrylamide gels to filters [5]. The transfer process takes place in low ionic strength buffers to prevent overheating and distortion of the gel. An external cooling system helps to minimize these effects. Large volumes of transfer solution are used to prevent the buffering capacity being exceeded. Home-made apparatus can be used, but commercial equipment is also available. Transfer is usually complete in under 3 h, but the time taken depends on the composition of the gel and the size of fragments being transferred.

Transfer can also be made from agarose gels, but there is a danger that the heat produced may cause the gel to melt. The problem can be alleviated by using an external cooling system.

10.2.3 Northern blots

Although RNA is single-stranded, it generally contains substantial stretches that are hydrogen bonded. For accurate determination of molecular weight, the RNA is denatured before applying to the gel and is kept denatured during electrophoresis. Pretreating the RNA with formamide /formaldehyde and running in a formaldehyde-containing gel is the most commonly used procedure [1]. An alternative method is to react the RNA with glyoxal/DMSO and to run the gel in neutral conditions. Glyoxal attaches covalently to the RNA and prevents renaturation, but it must be removed before hybridization as it inhibits duplex formation [6].

Northern blotting does not require pretreatment of the gel as the RNA is already denatured. RNA can be transferred by capillary action, vacuum blotting, positive pressure and electroblotting. It takes about 12 h to achieve efficient transfer by conventional capillary transfer. Downward capillary action in dilute alkali is quicker (about 1.5 h) and has the advantage that it removes glyoxal groups from the RNA. Vacuum blotting and positive pressure are also faster than conventional capillary methods and give sharper bands on hybridization.

10.2.4 Dot/slot blots

Dots were traditionally applied to filters by hand and for qualitative analyses this is satisfactory. No special equipment is required for manual application. A template can be drawn in pencil or stamped on to the membrane to guide the placement of dots. Commercial apparatus is available which uses a vacuum manifold to suck nucleic acid on to filters. It is usually supplied with interchangeable templates to produce either dots or slots of different dimensions. The dots have a more uniform diameter than those applied by hand, but slots may be preferred for quantitation as the nucleic acid seems to bind more evenly.

DNA is usually denatured before it is applied to the filter, but double-stranded DNA can be applied to nylon filters and denaturation can be carried out subsequently as for colony/phage treatment (Section 10.2.1). Both give satisfactory results, but denaturation after applying to the filter is easier.

RNA can be applied either by hand or via a manifold. The RNA is generally denatured with formamide/formaldehyde before being applied to the filter. Treatment with glyoxal before binding is also effective, but heat treatment followed by snap cooling in ice tends to give inadequate denaturation.

10.3 Immobilization of nucleic acid to filters

Nucleic acids transferred to nitrocellulose or nylon filters by any of the above methods are very loosely bound and would be readily lost from the filters if they were not attached more firmly.

10.3.1 Baking

Immobilization is traditionally achieved by baking in an oven for 2 hours at 80°C. A vacuum oven is recommended for baking nitrocellulose to reduce the possibility of explosion and fire, but is not necessary for nylon filters.

It is important to note that nucleic acid baked on to any type of filter is firmly, but not covalently bound. This has several consequences. First, nucleic acid is gradually leached off the surface when filters are incubated for long periods at high temperature. Second, if the probe in solution is complementary to the entire length of a filter-bound sequence, the hybrid will dissociate from the filter and will be lost in solution [7]. Both of these situations can cause loss of sensitivity. Nevertheless, baking is still widely used for immobilizing nucleic acid.

10.3.2 UV fixation

Covalent binding of nucleic acid to nylon filters can be achieved by UV irradiation of the filter. This is quick, taking only a few minutes and is probably much more efficient than baking particularly if used in conjunction with transfer in buffers of high ionic strength. UV irradiation links a small proportion of the thymine residues of the DNA to the filter. If the filter is over-irradiated, too many thymine residues will be linked and the DNA will lose its ability to hybridize, so the sensitivity will be reduced. If the filter is under-irradiated, the nucleic acid will dissociate from the filter and will be lost. So, to achieve maximum sensitivity, the irradiation step must be optimized.

The amount of irradiation required for optimal binding depends on the dampness of the filter. The drier it is, the less irradiation is required. So, when preparing membranes for irradiation, it is important to try and reproduce the same conditions each time.

Satisfactory binding can be obtained with UV transilluminators of the type used to visualize nucleic acids in ethidium bromide-treated gels. The damp filter is placed with the side containing the nucleic acid facing the UV source and irradiated. As the dose of irradiation emitted from UV tubes decreases with age and the amount of use, the illuminator must be calibrated regularly to determine the optimal time of exposure. This is easily done by preparing a nylon strip with replicate dots of denatured nucleic acid and exposing the dots for different lengths of time to the UV source. Hybridization is then carried out and the irradiation time giving the strongest signal is chosen for subsequent filters.

Calibrated UV cabinets are commercially available. The most useful emit a constant amount of energy (e.g. $120 \, \text{mJ} \, \text{cm}^{-2}$) automatically irrespective of the age of the UV lamp.

Covalent immobilization of nucleic acid to damp nitrocellulose by UV illumination has been reported, but is not recommended because of the danger of explosion and fire.

10.3.3 Alkali fixation

Capillary transfer of DNA or RNA in alkali immobilizes the nucleic acid covalently to positively charged nylon filters and some makes of ordinary nylon filters. This is very useful because self-complementary sequences do not have a chance to reanneal before they are bound. In addition small molecules are bound very efficiently because they are bound covalently as soon as they reach the filter and consequently have little time to diffuse away. Loss of small molecules can be a problem when neutral transfer solutions are used in capillary transfer. A disadvantage with binding in alkali is that background is often high on hybridization.

If nucleic acid has been immobilized by one method, it is unnecessary and even detrimental to use a second. For example, if DNA has been covalently bound to nylon membranes by alkali treatment , it should not also be treated with UV light as the hybridization signal will probably be reduced.

10.4 Hybridization in dried gels

Hybridization of oligonucleotide probes takes place efficiently in agarose gels that have been dried down after electrophoresis [8,9]. This procedure is particularly useful when the nucleic acid targets are small molecules because they tend to be lost by diffusion if transfer is carried out by capillary transfer. This is sometimes jokingly referred to as an 'unblot'.

10.5 Experimental procedures

10.5.1 Preventing degradation of RNA

Many hybridizations involve handling RNA. Ribonucleases are so widespread that it is necessary to take stringent precautions to prevent RNA being degraded. These are discussed before procedures for hybridization,

- A set of chemicals and stock solutions should be kept for RNA work only.
- Glassware should be kept separate from the rest of the laboratory glassware. It should be baked at 180°C for at least 8 h (overnight) or 250°C for at least 4 h.
- Distilled or de-ionized water. Add diethyl pyrocarbonate (DEPC) to 0.1%, shake and allow to stand for 2 h. DEPC is toxic, so use gloves and work in a fume hood. Destroy the DEPC by autoclaving otherwise it can react with the adenine in RNA and with primary amines such as Tris.
- Solutions can be prepared with untreated de-ionized or distilled water then treated with DEPC. However, for Tris-containing solutions, use DEPC-treated water and fresh chemical.
- Disposable plasticware such as tips for liquid handling and micro-centrifuge tubes can be used without treatment, but wear gloves and set aside unopened packages for RNA work only.
  For plastic beakers, etc. rinse in DEPC-containing water, then DEPC-treated water. Dry in air and autoclave.
- The skin is a good source of ribonuclease so plastic or latex gloves should be used at all times when handling RNA-containing solutions or the glassware, solutions etc. which are to come in contact with the RNA.

- RNA is degraded in solutions with a pH as low as 9, so keep the pH of buffers around neutral.
- For equipment such as gel electrophoresis tanks and dot-blot apparatus, try to keep one for RNA work only. Some makes of tank/dot-blot apparatus can be treated with dilute alkali then rinsed thoroughly in DEPC-treated water. Alternatively soak the tank/blot apparatus in 3% hydrogen peroxide for 10 min. Rinse thoroughly in DEPC-treated water.

However, it is important to check with the manufacturer's instructions for RNase inactivation.

10.5.2 Phage/colony transfer

PROTOCOL 10.1: Phage/colony transfer

1. Using flat-bladed forceps apply the membrane to the surface of a Petri dish. This is easily done for small plates by touching one edge of the filter to the surface of the agar and gradually lowering it until it lies evenly on the surface. For larger plates, it may be more convenient to hold the filter at opposite edges with forceps and allow the center to sag a little. Lower the filter to touch the center of the dish at one point and slowly lower each edge as the region of 'wetness' spreads over the filter from the region of original contact. Do not allow air bubbles to be trapped under the filter as they will prevent contact between the filter and the colonies on the surface of the agar.
2. Using a syringe with dye such as Indian ink pierce the surface of the filter in three asymmetric positions so that the filter can subsequently be aligned with the Petri dish. Some researchers prefer to make the alignment marks on the upper surface of the filter (the side not touching the agar) by colored pen before the filter is applied to the plate. This does not affect the sterility of the filter as it is the lower surface which touches the agar. The alignment marks can then be copied on to the foot or side of the dish by marker pen. This prevents perforation of the filter by the needle which can lead to subsequent tearing of the filter.
3. Leave the filter in contact with the agar surface for 1 min. Carefully lift one edge of the filter using forceps and peel the filter off in one slow, steady movement.
4. Lyse cells and denature the DNA by placing the filter, colony side upwards, on a dish containing 3MM filter paper (Whatman) saturated with 1.5 M NaCl, 0.5 M NaOH. Ensure that no air bubbles are trapped beneath. Leave 5 min.
5. Place the filter, colony side upwards, on a dry 3MM filter for about 5 s to remove excess alkali.
6. Place the filter on to 3MM paper saturated with 1.0 M Tris-HCl, pH 7.5, 1.5 M NaCl to neutralize the filter. Leave 5 min.

For noncovalent binding on nitrocellulose, nylon or charged nylon filters
7. Remove the filter on to dry 3 MM paper and leave at room temperature for about 15 min.
8. Sandwich between sheets of dry 3MM filter paper and bake in a vacuum oven for 2 h at 80°C.

For covalent binding on nylon or charged nylon filters
7. Transfer the filter to a dry 3MM filter for a few seconds to remove excess neutralization solution.
8. *Either:*
   Place the damp filter, colony side downwards, on a transilluminator surface that has been covered with a sheet of Saranwrap. Wearing protective UV goggles or a UV mask

illuminate the sheet with long wavelength UV light for a length of time that has been empirically determined.

or:
Place the filter in a UV cabinet and irradiate according to the manufacturer's instructions.
9. Allow the filter to dry in air.

Alternative for covalent binding to charged nylon filters
1–4. Carry out steps 1–4 as above, but prolong the exposure of the filter to 1.5 M NaCl, 0.5 M NaOH to 15 min.
5. Neutralize the filter by placing on 3MM paper saturated with 1.0 M Tris-HCl, pH 7.5, 1.5 M NaCl for 5 min.
6. Remove filter to dry 3 MM filter paper and air dry.

Notes
1. *For phage transfers*
   When the top phage-containing layer on the plate is agar, there is a tendency for some agar to adhere to the filter. This is liable to trap probe during hybridization which gives rise to high backgrounds. In addition, removal of agar is liable to include some plaques so that these phage are lost from the screening process as they can not be recovered. There are two ways to minimize the likelihood of agar sticking to filters.

   - Chill the plates at 4°C for 1 h before taking plaque lifts. This hardens the agar.
   - Or better, because agarose is less likely to be removed from the plate than agar, so use a top layer of 0.7% agarose instead of agar to spread the phage and bacterial host on the plate. Chill the plates at 4°C before applying the filter.

2. Several plaque lifts can be taken from the same plate while still retaining the ability to recover live phage from the remaining plaque. This is useful if different probes are to be used.
3. *For both plaque and colony lifts*
   It sometimes happens that when a filter is applied to a plate, it does not sit flat or it traps bubbles beneath part of the filter. This is more likely to occur with large rather than small Petri dishes. In this event do not adjust the position of the filter. This will cause smearing of the plaques or may even dislodge the agarose layer. Instead lift the filter smoothly and discard it after autoclaving. Try again with a new filter.

10.5.3 Southern blots

It will be assumed that DNA has been separated according to size on agarose gels using standard buffers such as Tris/borate/ethylenediaminetetraacetate (EDTA) or Tris/Acetate/EDTA containing 0.5 μg ethidium bromide ml^{-1} [1].

- Gels of between 0.5% and 1.2% agarose are commonly used for Southern blots. The lower the concentration of agarose, the more flexible the gel is and the more careful handling it requires.
- The DNA can be transferred through either the upper or lower surface of the gel. Some researchers prefer the lower surface because the DNA usually has a shorter distance to travel to reach the filter. This,

however, depends on the depth of the gel, the depth of sample-containing solution applied to the loading wells and how close the loading wells are to the lower surface.

- If the DNA sequences of interest are larger than about 8 kb, it may be necessary to reduce their size by depurination. This will depend on the concentration of agarose in the gel because the higher the percentage of agarose, the more difficult it is for long molecules to diffuse through. Do not overdo depurination or the size of DNA may be become so small that it fails to bind or hybridize effectively.

PROTOCOL 10.2: Southern blots

A Preparing the gel

1. Remove the gel and cut off the upper left hand corner as an orientation signal. Using a sharp scalpel, trim off a narrow strip all the way round the edges of the gel. This is because the agarose tends to be deeper at the edges so its removal allows the gel and filter to form better contact.

B. Depurination

(Carry out if size of DNA needs to be reduced otherwise omit this step and go straight to step C.1)

1. Place the gel in a flat dish containing sufficient 0.25 M HCl to cover it. The gel may float, but it can be kept covered with solution by rocking or agitating very gently.

2. Treat for 15 min at room temperature then pour off the acid. This can be done by placing a gloved hand over the gel to keep it from moving and gently tilting the dish until the acid has been completely removed.

3. Add fresh HCl and repeat the treatment for a further 15 min. During acid treatment any bromophenol blue marker in the gel will turn yellow.

4. Remove the acid as completely as possible and rinse the gel very briefly (a few seconds) in distilled water to remove residual acid from the dish. Drain well.

C. Denaturation

1. Add sufficient 0.5 M NaOH, 1.5 M NaCl to cover the gel and rock it gently for 20 min at room temperature.

2. Remove the alkali solution and replace it with fresh 0.5 M NaOH, 1.5 M NaCl. Again rock gently for another 20 min.

   If the gel has been treated in acid, the bromophenol blue marker should become blue again on treatment with alkali.

3. Remove the solution and rinse the gel very briefly in distilled water to remove residual alkali. Drain well.

D. Neutralization of the gel

1. Add sufficient 0.5 M Tris-HCl, pH 7.5, 1.5 M NaCl to cover the gel. Rock it gently at room temperature for 20 min.

2. Remove the solution and drain well.

3. Repeat the neutralization treatment.

E. Capillary transfer in high salt conditions

1. Have ready a nitrocellulose or nylon filter cut to the exact dimensions of the trimmed gel. Use blunt forceps and gloved hands to handle the filter.

2. Wet the filter by floating it in distilled water, then immerse it for 2–3 min. Blot it gently on 3MM paper and transfer it to a solution of 20 × SSC (transfer solution).

3. Into a flat dish, arrange a platform on which the gel is to be placed and add 20 × SSC to the dish to act as a reservoir. The platform can be very simple, e.g. it could be the

electrophoresis apparatus used to run the gel, or four silicone corks supporting a glass plate in a dish.

4. Cut two pieces of 3MM paper to act as wicks. They should be as wide as the platform and long enough to pass over the platform and dip into the transfer solution (see *Figure 10.1*). Wet the wicks by running them through the transfer solution and place them one on top of the other on the platform. Roll with a test tube/glass rod/pipette to remove any air bubbles trapped beneath. Ensure that the ends of the wick dip into the reservoir.

5. Place the gel (inverted if desired) on the wet 3MM paper and roll to remove air bubbles. Place Saran Wrap or any other type of film closely round the sides of the gel to act as a barrier between the 3MM paper and the absorbent paper which will eventually be placed on the gel. The absorbent paper sometimes droops and if it touches the reservoir or damp 3MM paper liquid will bypass the gel and pass through the paper instead. This will reduce the efficiency of transferring DNA to the filter enormously.

6. Place the nitrocellulose/nylon filter (with one corner cut off) on top of the gel with the cut corners on gel and filter aligned. Roll to remove trapped air bubbles.

7. On top place two sheets of 3MM paper that are the exact size of the gel and have been wet in transfer solution. Roll to remove air bubbles.

8. On top place two dry sheets of 3MM paper that are the exact size of the gel.

9. On top place a stack of dry, absorbent paper towels 5–10 cm high. Place a glass/perspex plate on top and a weight (about 1/2–1 kg). Alternatives to towels are paper napkins or the absorbent portion cut out of baby's disposable nappies/diapers. Two of the latter folded to fit are very absorbent and very cost effective.

10. Transfer of the DNA from the gel to the filter takes place as the solution moves gently upwards through the gel carrying the DNA with it. When the DNA reaches the filter which is on top of the gel, it becomes loosely bound.

11. Allow transfer to proceed for 8–16 h. The time required will depend on the depth of gel and size of DNA fragments to be transferred. Overnight transfer is often convenient.

12. Remove the towels and with a ball point pen, lightly mark the positions of the wells on the filter. Gently peel the filter off.

F. Immobilizing DNA on the filter

1. *Either* Baking
   - Place the filter on 3MM paper and allow to air dry for 30 min at room temperature.
   - Bake 80°C for 2 h (in a vacuum oven if appropriate).

   Or UV treatment
   either:
   Place the filter DNA side down on to Saran Wrap on a UV transilluminator. Illuminate for the time that has been optimized as described in Section 10.3.2.

   or:
   Place the filter in a commercial UV cabinet and irradiate according to the manufacturer's instructions.

G. Capillary transfer in alkali for positively charged nylon membranes

1. After running the gel, depurinate the DNA as in step *B*.1–4 above if necessary.

2. Denature the gel twice in 0.4 M NaOH, 1.0 M NaCl for 20 min each time.

3. Carry out capillary transfer as described in steps *E*.1–12 above, but using 0.4 M NaOH, 1.0 M NaCl as transfer solution. Avoid long exposure of charged nylon to alkali or high backgrounds may arise.

4. After transfer, rinse the filter in 0.5 M Tris-HCl, pH 7.5 for 15 min to neutralize the filter and remove adherent DNA.

5. Place the filter on 3MM paper and allow to dry for 30 min at room temperature. The filter requires no further treatment for immobilizing DNA.

Notes
1. Gels tend to become dryish and rubbery as capillary transfer proceeds and there is probably no advantage in blotting for longer than 3–4 h. However, for convenience, blotting is often carried out overnight.
2. After transfer, the gel should be stained to determine the efficiency of transfer of DNA. Place the gel in a solution containing 0.5 µg ml^{-1} ethidium bromide (e.g. the buffer used to run the gel). After 30–45 min any remaining DNA on the gel should be visible under UV light. Even if no DNA is left in the gel, it does not follow that it has all been bound to the filter. Small DNA molecules in particular are liable to diffuse laterally and may not reach the filter.
3. After the filter is removed from the gel there should be no agarose sticking to the filter. However, if there is, rinse the filter very briefly in 20 × SSC. Remove the filter and allow excess liquid to run off.

10.5.4 Northern blots

It will be assumed that RNA has been separated according to size on standard formaldehyde-containing agarose gels [1]. RNA gels often contain molecular weight markers, but the gels are run in the absence of nucleic acid stains such a ethidium bromide. For many years there has been debate as to whether staining the gel before blotting reduces the efficiency with which RNA is transferred to the filter. There are several ways of avoiding the problem altogether. One is to incorporate a lane with markers labeled with radioactivity or a nonradioactive reporter such as digoxygenin (DIG). These can be detected at the same time as the hybrids are detected. Alternatively to monitor migration, it is common to run an extra lane of RNA and to excise and stain it.

PROTOCOL 10.3: Northern blotting

Solutions
Transfer Solutions
10 × SSC   1.5 M NaCl, 0.015 M trisodium citrate, pH 7.0.
or
10 × SSPE   1.8 M NaCl, 0.1 M sodium phosphate, pH 7.4, 10 mM EDTA

Method
1. In a fume hood soak the gel in two changes of RNase-free water for 15 min each time. This removes some of the formaldehyde and improves the efficiency of transfer.
2. Wet nitrocellulose or nylon or charged nylon filter in water then in transfer solution. (Some makes of nylon filters can be wet immediately in transfer solution. Check manufacturer's advice.)
3. Transfer the RNA to the filter by the capillary method described for DNA blots (Protocol 10.2 steps E.1–12) but use one of the transfer solutions above.

Notes
1. The conditions under which the gel is run keep the RNA denatured. So there is no need to pretreat the gel prior to transfer.
2. Formaldehyde-containing gels are very floppy and require careful handling.
3. All reagents should be treated to prevent RNase contamination. This includes baking

glassware, treating aqueous solutions with DEPC and keeping a set of reagents specifically for RNA work [1].

4. Although RNA is degraded by alkali, partial hydrolysis can improve the transfer of high molecular weight RNA. After rinsing the gel in water to remove formaldehyde, the gel is treated with 50 mM NaOH, 10 mM NaCl for 45 min. The filter is rinsed in normal transfer solution before setting up the transfer.

5. Some brands of filter require transfer in low concentrations of salt for optimal binding. The manufacturer's recommended conditions should be followed closely.

6. To stain a strip of gel containing markers, wash the strip twice in water for 20 min each time then twice in 0.1 M ammonium acetate for 20 min each time. This removes formaldehyde which binds stain. Stain the strip in 0.1 M ammonium acetate containing 2 µg ethidium bromide ml^{-1} for 15 min. Destain twice in 0.1 M ammonium acetate for 25 min each time. Visualize the RNA under UV light on a standard transilluminator.

10.5.5 DNA dot/slot blots

There is much flexibility in the type of DNA that can be used. Genomic, plasmid, phage DNA and fragments of DNA can be screened on dots. The DNA need not be pure; aliquots of 'mini preps' as used for characterization of recombinant phage and plasmids can be used as can partially purified eukaryotic DNA. If supercoiled DNA is to be analyzed, it should be converted into open circular or linear form before binding to filters. This is because the DNA needs to be single-stranded to bind to filters and supercoiled DNA renatures so quickly when the denaturant is removed, that it can not be trapped in the denatured state.

There are many protocols for dot/slot blotting. DNA can either be denatured before or after it is applied to the filter. Both give satisfactory results, but denaturation after applying to the filter is easier and is described in *Protocol 10.4*.

PROTOCOL 10.4: DNA dot/slot blots

Filters are usually pretreated in a solution of high ionic strength because this increases binding efficiency and helps to keep the diameter of the dots small. The solutions usually used are 1 M ammonium acetate or 20 × SSC (1 × SSC is 0.15 M NaCl, 0.015 M trisodium citrate).

1. Wet nitrocellulose or nylon or charged nylon filter by floating it in a dish of water. When one side is wet, immerse the filter to wet the other side. Blot lightly with 3 MM Whatman paper to remove excess liquid.

Either transfer by hand

2. Transfer to a dish containing 20 × SSC or 1.0 M ammonium acetate and agitate gently for 5–10 min. Blot lightly and dry at room temperature or under a heat lamp.

3. Before applying dots by hand it may be convenient to mark out 5 mm circles into which the dots are placed. This can be achieved using a custom-made rubber stamp and ink pad. This allows the dots to be easily identified. Do not use much ink or all the solutions into which the filter is subsequently placed will become colored.
   When applying the DNA, do not allow the filter to touch any surface at the positions of the dot as this causes the DNA solution to diffuse. This can be prevented by balancing

the filter on the rim of a beaker or plastic box so that the bulk of the surface is not touching anything. In this way the DNA dries in air and maintains tight spots
4. Apply the DNA solution (in TE buffer, pH 8.0) to the circles on the filter using an automatic pipette. Do not allow the dots to spread. If necessary, make repeated applications of about 2 µl each time allowing each application to dry before applying the next.
5. Dry the samples in air or place under a heat lamp.

Or transfer by blotting apparatus
2. Transfer to a dish containing 20 × SSC or 1 M ammonium acetate and agitate gently for 5-10 min. Blot lightly, but do not allow to dry.
3. Cut Whatman 3MM paper to fit the manifold and wet it in 2 × SSC. Place on the lower part of the manifold and place the filter that has been treated with high salt buffer on top. Eliminate air bubbles between the two sheets.
4. Cover any manifold positions that will not be used with Parafilm/Nescofilm and fit the top section of the manifold in position.
5. Apply a gentle suction with a vacuum pump or a water pump with a trap. Test the strength of suction by placing about 500 µl 20 × SSC in a well and watching how quickly it disappears. Adjust the vacuum so that it takes about 5 min for 500 µl to flow through.
6. Turn off the vacuum and add 20 × SSC to the wells to keep the dot positions wet. If the filter is allowed to dry, the DNA solution can seep sideways through the foot of the well giving irregularly shaped dots.
7. When the DNA is ready, turn on the vacuum to drain the liquid from the wells. Turn off the vacuum and immediately apply the DNA solutions to the wells. Turn on the vacuum and allow the suction to draw the DNA on to the filter. The low rate of application is required to get efficient binding of DNA.
8. Remove the filter and allow the samples to dry in air or place under a heat lamp.

Fixing DNA to the filter
9. Place the filter, DNA side up, on 3MM paper saturated in 1.5 M NaCl, 0.5 M NaOH. Leave for 5 min. Place on dry 3MM paper for a few seconds to remove excess alkali.
10. Transfer the filter to Whatman 3MM paper saturated in 0.5 M Tris HCl, pH 7.5, for 30 s. Place on dry 3MM paper for a few seconds to remove excess solution.
11. Transfer the filter to Whatman 3MM paper saturated in 1.5 M NaCl, 0.5 M Tris-HCl, pH 7.5 for 5 min.
12. For nitrocellulose filters, place the filter on a dry sheet of 3 MM paper and air dry. Bake at 80°C for 2 h in a vacuum oven
   For nylon or charged nylon filters
   either
   place the filter on a dry sheet of 3 MM paper, air dry and bake at 80°C for 2 h *or* treat with UV for a time determined empirically (see Section 10.3.2) and air dry.

Alternatively for charged nylon filters
10. Treat in alkali for 15 min and omit steps 11 and 12 above.
11. Rinse the filter by immersing in 5 × SSC for about 1 min.
12. Place the filter on a dry sheet of 3 MM paper and air dry
13. The filters are now ready for prehybridization.

Notes
1. Up to about 50 µg DNA can be applied per well to nylon filters and about 10 µg per well for nitrocellulose, but it is seldom necessary to apply so much.
2. Try to keep the volume of DNA small (10 µl or less) when applying dots by hand so that the number of applications is kept small and the dots remain tight.
3. When transferring by blotting apparatus, apply the DNA in volumes of 100–200 µl. This is necessary to get even binding of the DNA.

10.5.6 *RNA dot/slot blots*

The principle of RNA dot blots is exactly the same as for DNA dot blots. Samples of RNA are applied to filters in an array, bound and hybridized with a labeled probe. The extent of hybridization can be estimated by comparison with suitable standards. The technique is sensitive with as little as ~1 pg being detectable. Nitrocellulose, nylon and charged nylon filters are all suitable for RNA dot blots, but the solutions required for binding may differ according to the type of filter. Consult the manufacturer's instructions. Probes can be labeled with radioactive or nonradioactive reporter molecules.

The RNA can be total or polyadenylated RNA. Although RNA is single-stranded, it contains stretches that are double-stranded and these must be denatured for efficient binding to the filter. Heat treatment does not give efficient binding, and alkali treatment is unsuitable since it degrades RNA. RNA denaturants such as methyl mercury are not used because of their toxicity. The two most commonly used methods of RNA denaturation are formaldehyde/formamide treatment (*Protocol 10.5*) and DMSO/glyoxal treatment (*Protocol 10.6*). Both methods give good results and it is a matter of personal preference which is used. Glyoxal binds covalently to guanine residues in DNA and RNA forming an adduct that is stable at acid and neutral pHs. The nucleic acid must be pure otherwise proteins will be bound to the nucleic acid. After binding the glyoxal groups must be removed as they inhibit hybridization. This is easily achieved by heating in mild alkaline conditions.

PROTOCOL 10.5: RNA dot blots – formaldehyde treatment
Solutions

10 × MOPS buffer   0.2 M 3-(N-Morpholino)-propane-sulfonic acid
                    50 mM sodium acetate, pH 7.0
                    10 mM EDTA
                    This buffer is light sensitive and has limited stability. It should be made up frequently and stored in the dark.

Formaldehyde denaturation   Mix together 700 μl formamide, 233 μl formaldehyde
solution   (13.2 M stock solution), and 70 μm 10 × MOPS.

Method

1. Pre-wet the filter in RNase-free water, then soak in 10 × SSC for 1 h.
   Blot on Whatman 3 MM paper to remove excess liquid, and dry in air or under a heat lamp.
2. Dissolve up to 10 μg RNA in 10 μl water or low salt buffer.
   Add 25 μl formamide denaturation buffer.
   Heat 65°C 10 min to denature the RNA. Snap cool in ice.

Either transfer by hand

3. Before applying dots by hand it may be convenient to mark out 5 mm circles into which the dots are placed. This can be achieved using a custom-made rubber stamp and ink

pad. This allows the dots to be easily identified. Do not use much ink or all the solutions into which the filter is subsequently placed will become colored.

4. When applying the RNA, do not allow the filter to touch a surface at the position of the dot as this causes the RNA solution to diffuse. To prevent this happening balance the filter on the rim of a beaker or plastic box so that the bulk of the surface is not touching anything. In this way the RNA dries in air and maintains tight dots.

5. Apply by hand in 2–3 µl aliquots drying *very* gently with *cold* air from a hair drier between applications. For amounts of RNA that can be applied, see note 2.

Or transfer by blotting apparatus

3. Cut Whatman 3MM paper to fit the manifold and wet it in 2 × SSC. Place on the lower part of the manifold and place the filter that has been treated with high salt buffer on top. Eliminate air bubbles between the two sheets.

4. Cover any manifold positions that will not be used with Parafilm/Nescofilm and fit the top section of the manifold in position. Clamp in position/fasten retaining screws.

5. Apply gentle suction with a vacuum pump. Test the strength of suction by placing about 500 µl 20 × SSC in a well and watching how quickly it disappears. Adjust the water pressure so that it takes about 5 min for 500 µl to flow through.

6. Turn off the vacuum and fill the wells with 20 × SSC to keep the dot positions wet.

7. Apply the RNA solutions to the wells and allow the suction to draw the RNA on to the filter. The low rate of application is required to get efficient binding of RNA.

8. Rinse with 2 × 0.5 ml 20 × SSC.

9. Switch off the vacuum and disassemble the apparatus. Remove the filter.

Fixing of RNA to filter

10. For nitrocellulose filters, place the filter on a dry sheet of 3 MM paper and air dry. Bake at 80°C for 2 h in a vacuum oven.

    For nylon or charged nylon filters, either place the filter on a dry sheet of 3 MM paper, air dry and bake at 80°C for 2 h.

    or

    treat with UV light for a time determined empirically (see Section 10.3.2) and air dry.

11. The filters are now ready for pre-hybridization.

The solution applied to the filter contains: 50% formamide, 2.2 M formaldehyde, 0.5 × MOPS, up to 250 µg RNA ml^{-1}.

Notes

1. Serial dilutions of RNA can be made in a solution composed of 500 µl formamide, 167 µl formaldehyde (13.2 M stock solution), 50 µl 10 × MOPS, 283 µl RNase-free water.

2. Up to 50 µg RNA can be applied per dot on a nylon filter or about 5-fold less on a nitrocellulose filter. About 50 µg total RNA or 10 µg polyadenylated RNA should give an adequate hybridization signal.

3. To minimize leaking between slots/dots, make sure that the nuts on the retaining screws are tightened to the same extent all round.

4. If a vacuum pump is not available, the RNA can be sucked on to the filter using water pressure from a tap. Such a system *must* incorporate a trap. It is generally more difficult to control the suction using water pressure.

5. If RNA dots are applied too quickly to the filter (i.e. if the suction is too fierce), halos rather than filled circles will appear on hybridization.

PROTOCOL 10.6: RNA dot blots: glyoxal denaturation
Solutions
Glyoxal denaturation   For each 100 μl denaturation solution, mix 34 μl deionized glyoxal,
solution   20 μl 0.1 M sodium phosphate, pH 6.5 and 46 μl RNase-free water.

Transfer by hand
1. To RNA (≤ 8 μg) in 2 μl water, add 2 μl glyoxal denaturation solution.
2. Incubate for 1 h at 50°C.
   Minimize evaporation by keeping the tube well submerged.
3. Spin briefly to bring contents to the foot of the tube.
4. Apply the samples to the filter in 2 μl aliquots.
   Dry in air between applications or apply cold air gently from a hair dryer.
5. For nitrocellulose filters, place the filter on a dry sheet of 3 MM paper and air dry. Bake
   at 80°C for 2 h in a vacuum oven
   For nylon or charged nylon filters, either place the filter on a dry sheet of 3 MM paper air
   dry and bake at 80°C for 2 h;
   or treat with UV light for a time determined empirically (see Section 10.3.2) and air dry.

Removing glyoxal group
6. Heat 200 μl of 20 mM Tris-HCl, pH 8.0, to 100°C. Remove from the heat and carefully
   immerse the filter. Allow to cool to room temperature

The composition of the denaturation solution is 1.0 M glyoxal, 10 mM sodium phosphate, pH
6.5 and up to 2.0 mg RNA ml^{-1}.

Notes
1. Glyoxal is readily oxidized to glyoxylic acid which will degrade RNA. De-ionize it with a
   mixed-bed resin [1] and store in 100 μl aliquots at −20°C. Do not re-use an aliquot once
   it has been opened.
2. DMSO can be included in the denaturation mixture at a final concentration of 50%. But should
   be omitted if nitrocellulose filters are used because DMSO dissolves nitrocellulose.
3. Serial dilutions can be prepared in 0.1% sodium dodecyl sulfate (SDS) so that each
   dilution is in a final volume of 4 μl.

References

1. **Sambrook, J., Fritsch, E.F. and Maniatis, T.** (1989) *Molecular Cloning: a Laboratory Manual.* Cold Spring Harbor Laboratory Press, Cold Spring Harbor, New York.
2. **Southern, E.M.** (1975) *J. Mol. Biol.* **98:** 503–517.
3. **Olszewska, E. and Jones, K.** (1988) *Trends Genet.* **4:** 92–94.
4. **Braman, J. and Dycaico, M.** (1989) *Strategies* **2:** 59.
5. **Bittner, M., Kupferer, P. and Morris, C.F.** (1980) *Anal. Biochem.* **102:** 459–471.
6. **Thomas, P.S.** (1983) *Methods Enzymol.* **100:** 255–266.
7. **Haas, M., Vogt, M. and Dulbecco, R.** (1972) *Proc. Natl Acad. Sci. USA* **69:** 2160–2164.
8. **Tsao, S.G.S., Brunk, C.F. and Pearlman, R.E.** (1983) *Anal. Biochem.* **131:** 3665–3672.
9. **Miyada, C.G. and Wallace, R.B.** (1987) *Methods Enzymol.* **154:** 94–107.

11 Basic techniques: labeling of probes

11.1 Introduction

Preparation of the probe usually involves three steps: incorporation of label, removal of nonincorporated precursors and measuring the efficiency of incorporation. The size of the probe can be controlled during labeling. Size is important because the rate of hybridization depends on the length of the probe (Section 3.3).

There are a variety of well-established methods for labeling probes with radioactive and nonradioactive reporter molecules (*Table 11.1*). Some methods incorporate the reporter molecule in an enzymatic reaction, others in a chemical reaction. Label can be incorporated uniformly throughout the probe or at the 3' or 5' ends. Commercial labeling kits are available which have been pre-optimized, are easy to use and give excellent results. Alternatively, labeling can be carried out with in-house reagents. Although it may be cheaper to buy individual reagents than a kit, it can be time-consuming to optimize reaction conditions.

11.2 Radioactive probes

For a radioactive probe to be useful for filter hybridizations, it must have a sufficiently high specific radioactivity, for example 2×10^7–2×10^9 c.p.m. ^{32}P μg^{-1} nucleic acid. The specific radioactivity attained depends on the amount of nucleic acid being labeled, the specific radioactivity of the radioactive dNTP used to label the probe, the concentration of unlabeled dNTP and the extent of incorporation of label into the probe. *Table 11.2* shows typical specific radioactivities that can be attained by incorporating ^{32}P into probes by different labeling

115

Table 11.1. Labeling of probes

Label	DNA uniform			DNA 5' end label	DNA uniform		DNA 3' end label/tailing Terminal transferase
	Nick translation	Random primer	Reverse transcription of RNA		Chemical	PCR	
Radioactivity	[α-³²P]dNTP	[α-³²P]dNTP	[α-³²P]dNTP	[γ-³²P]ATP	(Iodination)		[α-³²P]ddATP/dATP
Biotin	Biotin-dUTP	Biotin-dUTP	Biotin-dUTP	Linker required + activated biotin	Photoactivation	Biotin-dUTP	Biotin-ddUTP/-dUTP
Digoxygenin	DIG-dUTP	DIG-dUTP	DIG-dUTP	Linker required + activated DIG	Photoactivation	DIG-dUTP	DIG-ddUTP/-dUTP
Fluorescein	Fluorescein-dUTP	Fluorescein-dUTP	Fluorescein-dUTP	Linker required + activated fluorescein-NHS		Fluorescein-dUTP	Fluorescein-ddUTP/-dUTP
Enzymes				Linker required + modified enzyme	via glutaraldehyde		

	RNA uniform		Oligonucleotide			
	Run-off transcription	Photoactivation	5' end label kinase	3' end label terminal transferase	3' tailing terminal transferase	Internal
Radioactivity	[α-³²P]NTP		[γ-³²P]dATP	[α-³²P]ddATP	[α-³²P]dATP	
Biotin	Biotin-UTP	Biotin-aryl azide	Linker required + activated biotin	Biotin-ddUTP	Biotin-ddUTP	Biotin-phosphoramidites
Digoxygenin	DIG-UTP	DIG-aryl azide	Linker required + activated DIG	DIG-ddUTP	DIG-ddUTP	DIG-phosphoramidites or linker + activated ester of DIG
Fluorescein	Fluorescein-dUTP		Fluorescein-dUTP	Fluorescein-dUTP	Fluorescein-dUTP	
Enzymes			Linker required + modified enzyme			via glutaraldehyde

Table 11.2 Characteristics of ^{32}P-probes

Isotope	Half-life	Emission energy max. (Mev)	Emission energy mean (Mev)	Labeling methods	Typical sp. act. of probe (c.p.m. μg^{-1})
^{32}P	14.3 days	1.71	0.70	Nick translation	$1-5 \times 10^8$
				Random primer	$1-5 \times 10^9$
				Transcription	1×10^9
				PCR	$1-8 \times 10^9$
				End-labeling	$5 \times 10^6 - 10^9$
				Tailing	$\times 10^6 - 10^{10}$

methods. For most purposes, labeling with a single radioactive nucleotide gives an adequate specific radioactivity. But on rare occasions where a particularly high specific radioactivity is required, it may be necessary to incorporate two different radioactive nucleotides. It should be noted that maximal possible specific radioactivity is seldom desirable. The higher specific radioactivity a probe has, the more it is likely to undergo degradation because the sugar-phosphate backbone of the nucleic acid breaks at the position of ^{32}P decay and the radiation emitted may bombard the probe itself causing it to degrade.

DIG, biotin or fluorochrome is incorporated into probes by substituting about one-third of the TTP in the labeling reaction with Bio-, DIG- or fluorochrome-dUTP. The level of incorporation is about one modified nucleotide at every 20–25th position in uniformly labeled probes.

It is beyond the scope of this book to provide a comprehensive compendium of methods for labeling probes. Several of the most commonly used will be discussed and protocols given to allow a beginner to get started. A greater selection of protocols can be found in the literature [1,2].

11.3. Uniform labeling

11.3.1 Labeling of DNA by random priming

The most common method for uniform labeling of DNA is random priming (*Figure 11.1*). DNA is denatured and incubated with short primers of random sequence which anneal to the DNA. The primers are extended by the Klenow fragment of DNA polymerase I in the presence of labeled dNTPs to create labeled products. The method is very efficient and usually gives rise to net synthesis of DNA.

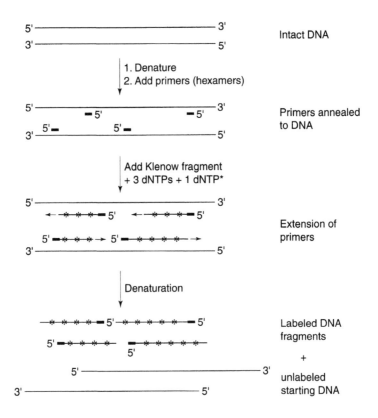

Figure 11.1. Uniform labeling of DNA by random priming. DNA is denatured and re-annealed in the presence of random hexamers which act as primers for the Klenow fragment of DNA polymerase I. In the presence of dNTPs a complement to the template is synthesized. By adding a labeled dNTP, uniformly labeled DNA is produced.

The final specific activity of the probe depends both on the ratio of unlabeled:labeled dNTP and on the length of time for which labeling occurs. Specific activities in excess of 2×10^9 c.p.m. μg^{-1} DNA can be reached. The length of the product is controlled by the ratio of primer:Klenow fragment of DNA polymerase I. The method in *Protocol 11.1* has been adapted from the original method of Feinberg and Vogelstein [3].

PROTOCOL 11.1: Labelling DNA by random priming

Solutions

- 5 × Reaction buffer   250 mM Tris-HCl, pH 7.2
  50 mM MgCl$_2$
  10 mM DTT
- Primers   Random hexamers at 1.56 μg ml^{-1}
- 10 × dNTP mix for radiolabeling   200 μM dATP, 200 μM dGTP,
  200 μM TTP, pH 6.5

or
- 10 × dNTP mix for nonradiolabeling
- Nucleotide labeling solutions

2 mM each of dATP, dGTP, dCTP, pH 6.5
[α-^{32}P]dCTP 3000 Ci mmol^{-1} (1.11 × 10^{14} Bq mmol^{-1}); 10 mCi ml^{-1} (3.7 × 10^8 Bq ml^{-1})
or DIG/TTP mix:
0.35 mM digoxygenin-11-dUTP
0.65 mM TTP
or biotin/TTP mix:
0.35 mM biotin-16-dUTP
0.65 mM TTP
or fluorescein/TTP mix:
0.5 mM fluorescein-12-dUTP
0.5 mM TTP

- Linear DNA
  (freshly diluted)
- Klenow fragment of DNA polymerase I
- 10 × Stop mix

1–25 µg ml^{-1} in TE buffer pH 8.0,
5 units µl^{-1}
100 mM EDTA, 1% SDS

Method
1. Mix:
- Linearized DNA   (up to 25 ng)
- Primers   5 µl
- Make up the volume to 34 µl with distilled water
2. Denature in boiling water, 4 min. Snap cool in ice (see note 2)
3. Spin briefly to move the contents to the foot of the tube
4. Add to annealed primer:
- 5 × Reaction buffer   5 µl
- 10 × dNTP mix for radiolabeling   5 µl
- [α-^{32}P]dCTP   5 µl
or
- 10 × dNTP mix for nonradiolabeling   5 µl
  and
- nucleotide labeling mix as appropriate   5 µl
- Klenow fragment of DNA polymerase I   1 µl
  Total volume   50 µl
5. Incubate at 37°C for 60 min
6. Add 5 µl Stop mix

The concentration of reactants for radioactive labeling is 50 mM Tris-HCl, pH 7.2, 10 mM MgCl$_2$, 2 mM DTT, 20 µM each dATP, dGTP, TTP; 0.33 µM [α-^{32}P]dCTP, up to 500 ng DNA ml^{-1}, 156 µg primers ml^{-1}, 100 U Klenow polymerase ml^{-1}.
For nonradioactive labeling the concentration of dNTPs is 200 µM each dATP, dCTP, dGTP; 65 µM TTP and either 35 µM DIG-dUTP *or* 35 µM biotin-dUTP *or* 35 µM fluorescein-dUTP.

Notes
1. DNA from molten agarose can be labeled without first removing the agarose, but the specific activity achieved is slightly less [4].
2. If the DNA is in a molten gel slice, cool rapidly to 37°C instead of placing in ice.
3. The above reaction requires 60 min incubation at 37°C for optimal incorporation. If smaller amounts of DNA are used, the reaction occurs more slowly and may take up to 5 h. An alternative is to incubate at room temperature overnight.
4. If more DNA is to be labeled, then the whole reaction should be scaled up. Higher concentrations of DNA in 50 µl lead to lower specific activities and shorter average probe lengths.
5. T7 DNA polymerase can be used in place of Klenow enzyme. Biotinylated nucleotides are incorporated more efficiently using nonamer primers and T7 DNA polymerase.

11.3.2 *Labeling of DNA by nick translation*

DNA can be uniformly labeled by nick translation. This procedure depends on the concerted action of two enzymes, DNase I and *E. coli* DNA polymerase I (*Figure 11.2*). DNase I introduces single-stranded nicks into a double-stranded DNA template at a frequency that depends on the amount of enzyme present. Starting at a nick, the $5' \rightarrow 3'$ exonuclease activity of *E. coli* DNA polymerase I excises nucleotides from the newly created $5'$ end. At the same time the $5' \rightarrow 3'$ polymerase activity adds template-specified nucleotides at the $3'$ end. The net result is that the nick is moved (translated) in the $5' \rightarrow 3'$ direction as new nucleotides are added in the same direction. If one of the added nucleoside triphosphates is labeled, a uniformly labeled product is generated. The proportion of bases replaced by DNA polymerase I and hence the specific activity of the product depends on the number of nicks introduced into the DNA. The final length of probe is determined by the ratio of DNase I : DNA polymerase I.

Nick translation used to be the most common method for labeling probes until random priming was devised. The advantage of random priming is that it tends to give higher specific activity of probes than nick translation, but nick translation gives better control of the length of the probe. *Protocol 11.2* was originally described by Rigby [5] and later modified for nonradioactive labels [6].

11.3.3 *Uniform labeling of RNA*

RNA probes are usually obtained by run-off transcription of cloned DNA from vectors that carry phage promoters (*Figure 11.3*). Several vectors are available. Among the most versatile are those that carry two different promoters which are orientated toward each other and separated by the multiple cloning site. The DNA is cloned between the promoters. This allows either sense or antisense RNA to be transcribed from the same plasmid and means that the direction of cloning of the insert is irrelevant for generating probes. Before transcription, the plasmid is linearized by exhaustive digestion with a restriction enzyme which cuts within or immediately adjacent to the insert. This is to prevent transcription continuing into the vector. See *Protocol 11.3*.

An alternative to using linearized plasmids as templates, is to amplify the cloned DNA by PCR using primers that carry appropriate phage promoters. After amplification, the phage RNA polymerase is used in a second reaction to generate single-stranded RNA [7,8].

Transcription is in the presence of a labeled NTP and gives rise to a product with high specific activity. The procedure is very efficient and it

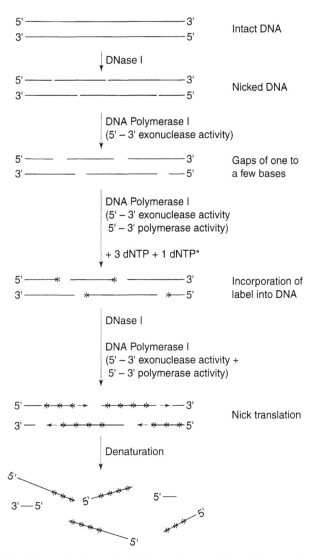

Figure 11.2. Uniform labeling of DNA by nick translation. Pancreatic DNAse I introduces a single-stranded nick into DNA at a frequency that depends on the concentration of enzyme. The 5′→3′ exonuclease activity of DNA polymerase I excizes nucleotides sequentially from 5′ end of the nick whilst the 5′→3′ polymerase activity of the same enzyme adds new nucleotides to the exposed 3′ termini. By adding a labeled dNTP, label is introduced into the DNA.

PROTOCOL 11.2: Labeling DNA by nick translation

Solutions

- 5 × Reaction buffer
  250 mM Tris-HCl, pH 7.8
  25 mM MgCl$_2$
  5 mM DTT
  250 μg ml^{-1} bovine serum albumin

- 5 × dNTP mix for radiolabeling

  1 mM each of dATP, dGTP, TTP, pH 6.5

  or

  5 × dNTP mix for nonradiolabeling

  0.25 mM each of dATP, dGTP, dCTP, pH 6.5

- Nucleotide labeling solutions

  [α-^{32}P]dCTP 3000 Ci mmol^{-1} (1.11 × 10^{14} Bq mmol^{-1}); 10 mCi ml^{-1} (3.7 × 10^8 Bq ml^{-1})

  or DIG/TTP mix:
  0.08 mM digoxygenin-11-dUTP
  0.17 mM TTP

  or biotin/dTTP mix:
  0.08 mM biotin-16-dUTP
  0.17 mM TTP

  or fluorescein/dTTP mix:
  0.08 mM fluorescein-12-dUTP
  0.17 mM TTP

- DNA

  200 µg ml^{-1} in TE buffer

- Pancreatic DNase I

  1 mg/ml. Before use, dilute 1:10 000 fold to 100 pg µl^{-1} in 10 mM Tris-HCl, pH 7.4, 5 mM MgCl$_2$, 1 mg ml^{-1} albumin Pentax fraction V, 50% gycerol. Diluted enzyme can be stored at −20°C.

- DNA polymerase I

  5 U µl^{-1}

- 10 × Stop mix

  100 mM EDTA, 1% SDS

Method

1. Mix:

- Water   14 µl
- 5 × Reaction buffer   10 µl
- 5 × dNTP mix for radiolabeling   10 µl

  or

  5 × dNTP mix for nonradiolabeling   10 µl
- DNA   5 µl
- Nucleotide labeling solution as appropriate   5 µl
- Diluted DNase I (100 pg/µl)   5 µl
- DNA polymerase I   1 µl
  Total volume   50 µl

2. Incubate at 14–16°C for 1–1.5 h.
3. To stop the reaction, add 10 × Stop mix   5 µl

The reaction components are: 50 mM Tris-HCl, pH 7.18, 5 mM MgCl$_2$, 1 mM DTT, 50 µg BSA ml^{-1}, 200 µM unlabeled dNTPs, 0.33 µM [α-^{32}P]dCTP (or 50 µM each dATP, dCTP, dGTP; 34 µM TTP and 16 µM nonradioactively labeled dUTP), 20 µg DNA ml^{-1}, 10 ng DNase 1 ml^{-1} and 100 U DNA polymerase I ml^{-1}.

Notes

1. Do not increase the incubation temperature above 16°C as template switching to the newly synthesized strand may occur [1] giving rise to hairpin or 'snapback' structures.
2. Do not incubate for longer than 1.5 h as after this time DNase I and exonuclease activities become relatively more important than polymerase activity and the specific activity will be reduced.
3. The method is sensitive to impurities such as traces of agarose, so gel-derived DNA fragments must be carefully purified.
4. Different batches of enzymes vary in their activities and contaminating DNases so the appropriate concentrations of DNase I and DNA polymerase I should be titrated. A mixture of the appropriate amounts of enzymes can be prepared and stored at −20°C.

To do this, set up several nick translation reactions all with the same amount of DNA polymerase I (say 5 Richardson units) and different amounts of freshly diluted DNase I (covering the range 0.01–0.1 ng). The size of products can be measured on alkaline agarose gels and incorporation of isotope by precipitation with trichloroacetic acid. Choose a concentration of DNase I that gives about 30% incorporation of radionucleotide and a product length in the range 400–800 nucleotides.

Pretitrated mixtures are available in commercial kits. Note that the stabilities of the two enzymes are not equal so it is best to use mixtures relatively quickly.

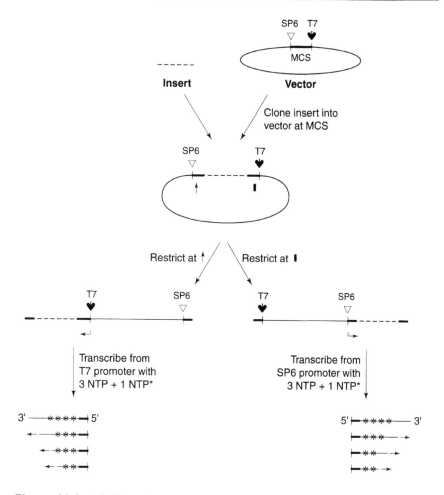

Figure 11.3. Labeling of RNA probes. A DNA fragment is cloned into a vector containing promoters for bacteriophages T3/T7 or SP6. The recombinant plasmid is linearized with a restriction enzyme to limit the size of transcript subsequently produced. RNA is synthesized from the bacteriophage promoter using bacteriophage RNA polymerase and NTPs, at least one of which is labeled. The sense of the RNA transcript depends on the end of the cloned insert from which transcription is initiated.

PROTOCOL 11.3: RNA labeling

Solutions

• 10 × transcription buffer	400 mM Tris-HCl, pH 7.5
	60 mM MgCl$_2$
	20 mM spermidine
	100 mM dithiothreitol
	1 mg ml^{-1} bovine serum albumin fraction V (Sigma)
• 10 × NTP mix for radioactive labeling	10 mM each of ATP, UTP, GTP
or 10 x NTP mix for nonradioactive labeling	10 mM each of ATP, CTP, GTP
• Nucleotide labeling solutions	[α-^{32}P]CTP
	400 Ci mmol^{-1} (1.48 × 10^{11} Bq mol^{-1});
	10 mCi ml^{-1} (3.7 × 10^8 Bq ml^{-1})
	or DIG/UTP mix:
	3.5 mM digoxygenin-11-UTP
	6.5 mM UTP
	or biotin/UTP mix:
	3.5 mM biotin-16-UTP
	6.5 mM UTP
	or fluorescein/UTP mix:
	3.5 mM fluorescein-12-UTP
	6.5 mM UTP
• Linearized template DNA in TE buffer, pH 8.0	1 µg µl^{-1}
• T7, T3 *or* SP6 RNA polymerase	The activity of enzymes varies between companies.
• RNase inhibitor	10 U RNasin µl^{-1}
• Stop mix	100 mM EDTA, 1% SDS

Method

A. Reaction

1. Mix in the order shown and at room temperature:

• Water (RNase-free)	x µl (to make final volume to 20 µl)
• 10 x transcription buffer	2 µl
• 10 x NTP mix for radioactive labeling	2 µl
or	
10 x NTP mix for nonradioactive labeling	2 µl
• RNase inhibitor	2 µl
• Linearized DNA	1 µl
• [α-^{32}P]CTP	5 µl
or DIG/UTP mix	2 µl
or biotin/UTP mix	2 µl
or fluorescein/UTP mix	2 µl
• Enzyme	2 µl (use about 10 units)
Final volume	20 µl

2. Incubate at 37°C, 1–1.5 h for T7 or T3 polymerases
         or 40°C, 1–1.5 h for SP6 polymerase

B. To stop the reaction

• Add 10 × Stop mix	5 µl

C. To remove the DNA template

1. Add RNase-free DNase I	10 units

2. Incubate 37°C, 30 min

D. To reduce the size of RNA transcript
1. Add:
- Water   60 µl
- 0.4 M NaHCO$_3$ 20 µl
- 0.6 M Na$_2$CO$_3$ 20 µl
2. Incubate 60°C for required time (see notes)
3. Neutralize by adding:
- 1.3 µl glacial acetic acid
- 20 µl 3M sodium acetate, pH 5.2
4. Mix well and add 0.5 ml ethanol. Mix.
5. Chill in dry ice for at least 1 h and process as in *Protocol 11.7*

The components for labeling conditions are 40 mM Tris-HCl, pH 7.5, 6 mM MgCl$_2$, 10 mM DTT, 2 mM spermidine, 100 µg BSA ml^{-1}, 1 mM each unlabeled NTP and either 62.5 µM ^{32}P-labeled NTP, or, 650 µM UTP and 350 µM nonradioactively labeled NTP), 50 µg DNA ml^{-1}, 1000 units RNase inhibitor ml^{-1}, 500 enzyme units ml^{-1}.

Notes
1. Assemble at room temperature: otherwise, at 0°C, the spermidine in the reaction buffer will cause the DNA to precipitate.
2. Longer incubations do not increase the yield of labeled RNA. If larger amounts of RNA are needed, scale up all the reaction components in proportion, i.e. do not just add extra DNA template.
3. For SP6 polymerase, the rate of transcription is maximal at 40°C, but the proportion of full length transcripts is greater at 30°C [9,10].
4. Do not reduce the concentration of unlabeled NTPs in an attempt to obtain higher specific activity probes. The unlabeled NTPs will become limiting and incomplete transcripts will accumulate.
5. For radioactive labeling, [α-^{32}P]UTP can be used in place of [α-^{32}P]CTP (with unlabeled GTP, ATP and CTP).
6. The time for incubating in dilute alkali depends on the initial size of transcript and the desired final size. The incubation time can be derived from the relationship [2]:

$$\text{Time (min)} = \frac{L_o - L_f}{k\,(L_o)(L_f)}$$

where L_o and L_f are the lengths in kb of the primary and desired transcript, respectively and k is the rate constant. $k = 0.11$ cuts kb^{-1} min^{-1}.

is possible to synthesize up to 20 times by weight of the amount of DNA added. The exact amount synthesized will depend on the amount and purity of template and its size. The template DNA can be excised by digestion with RNase-free DNase.

There are several ways of controlling the size of the probe. The first is by the choice of the restriction enzyme site that is cleaved to prevent transcription going right round the plasmid. The second is to heat the probe in the presence of Mg^{2+} ions or thirdly to treat with dilute alkali.

11.3.4 Uniform labeling by chemical methods

Enzymes are attached to the probe by chemical means. The method is easy and efficient [1]. Photoactivatable biotin and digoxygenin can be used to label both DNA and RNA uniformly [2]. The attached groups are stable at alkaline pH and at high temperatures and can withstand UV irradiation. The size of nucleic acid is unchanged by the labeling procedure which is useful if the labeled products are to be used as molecular weight markers, but means that if the size of the probe is to be reduced, it must be done before rather than during labeling.

11.4 End labeling of nucleic acids

For nonradioactive labeling of oligonucleotides, label can be incorporated at either end during synthesis. If an unlabeled oligonucleotide is already available, it can be labeled at the 3′ end by the methods described above.

11.4.1 Labeling at the 5′ end

DNA, RNA and oligonucleotides can be radiolabeled at the 5′ ends by T4 polynucleotide kinase which catalyzes the transfer of the γ-phosphate from [γ-^{32}P]ATP to a free 5′ hydroxyl group. Chemically synthesized oligonucleotides have a free 5′ hydroxyl group and radioactive phosphate can be attached directly with no pretreatment of the oligonucleotide. The reaction is very efficient and high specific activities of ~10^9 c.p.m. μg^{-1} oligonucleotide are easily achieved.

DNA fragments derived by restriction enzyme digestion usually carry 5′ phosphates which must be removed with phosphatase to generate free 5′ hydroxyl groups. The phosphatase must then be completely inactivated or removed otherwise the [^{32}P]phosphate attached by the kinase will be removed by residual phosphatase activity. Calf intestinal alkaline phosphatase (CIAP) is used in preference to calf thymus phosphatase because it is easier to inactivate. Labeling with T4 polynucleotide kinase is most efficient with single-stranded nucleic acids or double-stranded molecules with 5′ overhangs.

There are several ways in which an oligonucleotide can be labeled with nonradioactive labels.

- Reporter molecules can be incorporated at the 5′ end of oligonucleotides at the time of synthesis via the appropriate phosphoramidites.
- Alternatively the oligonucleotide can be synthesized with an amino group at the 5′ end which acts as a linker for subsequent

PROTOCOL 11.4: Labeling at the 5′ end using T4 polynucleotide kinase

A. Dephosphorylation of 5′ ends

Solutions

- 5 × CIAP buffer
  100 mM Tris-HCl, pH 8.3
  5 mM MgCl$_2$
  0.5 mM ZnCl$_2$
- DNA 1–10 pmol ends (note 1) in 1–7 µl TE buffer
- Calf intestinal phosphatase
  0.5 U^{-1}µl

Method

1. Mix
- Water
  to give final volume of 10 µl
- 5 × CIAP buffer
  2 µl
- DNA
  up to 7 µl
- Calf intestinal phosphatase
  1 µl
  Final volume
  10 µl
2. If 5′ termini are protruding, incubate at 37°C for 30 min.
   Add another 0.1 units enzyme and incubate for a further 30 min at 37°C.
   If the 5′ ends are blunt or recessed incubate at 37°C for 15 min then at 56°C for 15 min.
   Add a further 0.5 units enzyme and incubate at 37°C for 15 min then at 56°C for 15 min.
3. Heat inactivate at 75°C, for 10 min.
4. Extract with phenol and ethanol precipitate (note 2).
5. Collect precipitate, wash and dissolve in TE buffer, pH 7.5 at 2–10 pmol ends µl^{-1} (note 2).

B. 5′ end labeling using T4 polynucleotide kinase

Solutions

- *Either* deprotected oligonucleotide in
  TE buffer
  2–10 pmol ends µl^{-1}
  or dephosphorylated DNA
  2–10 pmol ends µl^{-1}
- 10 × kinase buffer
  500 mM Tris-HCl, pH 7.5
  100 mM MgCl$_2$
  50 mM dithiothreitol (DTT)
  1 mM spermidine HCl
- [γ-^{32}P] ATP
  3000 Ci mmol^{-1} (1.11 × 10^{14} Bq mmol^{-1});
  10 mCi ml^{-1} (3.7 × 10^{8} Bq ml^{-1})
- T4 polynucleotide kinase
  10 units µl^{-1}
- Stop mix
  100 mM EDTA, 1% SDS

Method

1. Assemble at room temperature
- water
  9–13 µl
- 10 × kinase buffer
  2 µl
- oligonucleotide *or*
  1–5 µl (to give 10 pmol ends)
  dephosphorylated DNA
- [γ-^{32}P]ATP
  3 µl (10 pmol)
- T4 polynucleotide kinase
  1 µl
  Total volume
  20 µl
2. Mix and incubate at 37°C for 45 min.
3. Stop the reaction by adding 2 µl stop mix.

The reaction components for dephosphorylation are: 20 mM Tris-HCl, pH 8.3, 1 mM MgCl$_2$, 0.1 mM ZnCl$_2$, 0.1–1 pmol DNA ends µl^{-1}, 0.06–0.1 U CIAP µl^{-1}.
The reaction components for 5′ end-labelling are: 50 mM Tris-HCl, pH 7.5, 10 mM MgCl$_2$, 5 mM DTT, 0.1 mM spermidine HCl, 500 pmol DNA (or oligonucleotide) ends ml^{-1}, 500 pmol [γ-^{32}P]ATP ml^{-1}, 500 U T4 polynucleotide kinase ml^{-1}.

Notes
1. To calculate the number of pmol ends see Appendix B.
2. Phosphate and ammonium ions are potent inhibitors of T4 DNA kinase. So DNA should not be in or precipitated from solutions containing either ion.
3. The reaction contains equal concentrations of DNA ends and [γ-^{32}P]dATP. About 50% label will be transferred.
4. To label different amounts of DNA ends to the same specific activity, scale up or down the whole reaction mix so that the concentration of reactants remains constant.

attachment of the reporter molecule. The modified oligonucleotide is reacted with an amino-reactive ester of the reporter such as *N*-hydroxysuccinimidyl (NHS)-DIG to generate a 5' DIG-linked oligonucleotide.

- An unmodified oligonucleotide can be labeled at the 5' end in a two-step process. In step one, a thiophosphate group is added to the 5' end of the oligonucleotide by reaction with γ-S-ATP and T4 polynucleotide kinase. In step two, the modified oligonucleotide reacts with a thiol-reactive maleimide-reporter molecule which causes the reporter to be covalently linked to the oligonucleotide.

Protocols for 5' labeling can be found in the literature [2, 11]. Commercial kits are also available.

11.4.2 Labeling at the 3' end

Labeling DNA or oligonucleotides at the 3' end depends on the ability of the enzyme terminal transferase to add nucleotides to the free 3' hydroxyl in a template-independent manner. Short, labeled, homopolymer tails can be added by incubating the enzyme and DNA with a labeled reporter-dNTP (usually dATP). This is called DNA tailing. Addition can be restricted to a single nucleotide if the enzyme is provided with a reporter-ddNTP. Since dideoxynucleotides lack 2' and 3' hydroxyl groups, no further addition of bases can be made. This type of reactions is termed end-labeling. Commercial kits are available for labeling DNA at the 3' end and protocols can be found in references 1 and 2.

A useful feature of tailing is that it provides several reporter molecules per DNA or oligonucleotide molecule. This makes detection more sensitive than when a single reporter molecule is present. However, care must be taken in tailing that the number of adenylate residues added is not too great or the tail will be longer than the oligonucleotide itself and will modify its hybridization properties. When hybridizing with tailed probes, it is usually advisable to include poly(A) in the prehybridization and hybridization solutions to prevent the tail from hybridizing to T-rich regions of the target nucleic acid. A disadvantage of using tailed probes is that background is sometimes a problem.

11.5 Removal of nonincorporated nucleotides

After the probe has been labeled, it is usual to remove nonincorporated radioactive label because during hybridization free labeled nucleotides may bind nonspecifically to the filter causing high backgrounds. Removal also facilitates determination of the efficiency with which label was introduced into the nucleic acid and reduces the hazard of exposure to radiation.

However, removal of radioactive label may not be necessary if more than about 70% has been incorporated into the probe [1]. Nonradioactive labeled nucleotides tend to have very low affinity for the commonly used types of filter and so nonincorporated label is seldom removed.

There are several standard methods for removing nonincorporated nucleotides.

11.5.1 *Size exclusion*

Size exclusion chromatography on Sephadex® G-50 columns will remove unincorporated dNTPs and very short oligonucleotides. Columns are easily set up in siliconized 15 cm Pasteur pipets plugged with siliconized glass wool (*Protocol 11.5*).

Similarly, Sephacryl® S-400 Matrix columns can be used to separate labeled DNA comprising ⩾ 270 bases from smaller fragments and columns of Bio-Gel P-60 can be used to remove unincorporated nucleotides from oligonucleotides.

11.5.2 *Sephadex spin columns*

These columns are simple, quick and reliable to use. They are available commercially, but it is easy to set one up using a 1 ml disposable syringe plugged with siliconized glass wool (*Protocol 11.6*).

11.5.3 *Ethanol precipitation*

This method (*Protocol 11.7*) depends on the precipitation of nucleic acids greater than about 20 bases in length by ethanol in the presence of high concentration of salt. Free dNTPs remain in solution. The recovery of nucleic acid depends on the concentration of DNA and the fragment size, but is generally greater than 90%.

PROTOCOL 11.5: Size exclusion on Sephadex® G-50 columns

1. Set up a column in a 15 cm pasteur pipet and wash with TE buffer, pH 8.0.
2. Load the sample and allow it to wash into the column.
3. Add one drop of a mixture of 2% dextran blue, 0.25% phenol red dyes. The labeled DNA will elute just ahead of the dextran blue dye. The free nucleotides will elute just ahead of the phenol red dye.
4. Elute the column with TE buffer and collect five-drop fractions until the red dye elutes.
5. For ^{32}P-labeled probes, a hand-held monitor can be used to follow elution of the probe. Nonradioactive probes can be identified by spotting an aliquot of eluate (5–10 µl) on to a nylon filter and visualizing as in Section 11.3.2.

PROTOCOL 11.6: Size exclusion on Sephadex® spin columns

1. Add a slurry of Sephadex® G-50 in TE buffer, pH 8.0 to a plugged 1 ml syringe.
2. Place in a 15 ml centrifuge tube and centrifuge at 1600 *g* for 5 min. Add more slurry and repeat the centrifugation.
3. Repeat the procedure until the level of Sephadex is about 0.5 cm from the top of the syringe. Remove excess TE buffer at the top of the syringe.
4. Place a microcentrifuge tube into the base of a fresh centrifuge tube. The eluate will be spun into this microcentrifuge tube.
5. Dilute the labeling mix to 50–100 µl with TE buffer and add to the top of the Sephadex column.
6. Centrifuge at 1600 *g* for 5 min.

The nucleic acid should be eluted into the microcentrifuge tube in 50–100 µl.

11.5.4 Adsorption to silica particles

In the presence of a high concentration of chaotropic salt, nucleic acids adsorb to silica particles whereas proteins and oligonucleotides do not. The nucleic acid needs to be about 100 bases long to bind. The particles with nucleic acid attached are washed in high salt buffer then the DNA or RNA is eluted in a small volume of low (no) salt buffer such as TE buffer. The method is fast, yields are excellent, the nucleic acid is pure and the low salt elution buffer makes it very easy to adjust the ionic strength later. Kits with silica particles are commercially and very simple to use. The method can be used to purify both radioactive and nonradioactively labeled nucleic acids.

An alternative and cheaper method of purifying nucleic acid is to use glass particles [12].

11.5.5 Phenol extraction

Phenol extraction can be used to separate the probe from unincorporated precursors, but is best avoided when labeling with nonradioactive

---

PROTOCOL 11.7: Ethanol precipitation

1. To the labeled nucleic acid, add 0.5 vol. of 7.5 M ammonium acetate and mix.
2. Add 2.2–3 vol. of ice-cold ethanol (1 vol. = volume of nucleic acid + volume of ammonium acetate).
3. Mix and place at $-20°C$ or $-70°C$ or in solid CO_2 for at least 1 h.
4. Centrifuge at 12 000 g for 15 min in a microcentrifuge at 4°C. Remove and discard supernatant.
5. Resuspend the pellet in 10–50 µl TE buffer.

Notes

1. If the amount of nucleic acid present is less than about 10 µg ml^{-1}, it may be necessary to add a carrier to aid precipitation. After stopping the reaction, add 1 µl glycogen (20 µg µl^{-1}). Mix and carry out steps 1–5 as usual.
2. In step 2 an alternative to adding ammonium acetate for precipitating the DNA, is to add 0.1 vol. 4 M LiCl, mix and add 3 vols ice cold ethanol (1 vol. = volume of nucleic acid + volume of lithium chloride).
3. Two successive precipitations can be carried out to improve the removal of unincorporated nucleotides, but is seldom necessary.

PROTOCL 11.8: Purification of DNA on silica particles

Preparation of silica particles
1. Add 10 g silica to 100 ml phosphate-buffered saline. Mix and leave for 2 h to settle.
2. Remove the supernatant containing fines.
3. Repeat steps 1 and 2.
4. Centrifuge at 2000 g for 2 min.
5. Resuspend the pellet in 3 M NaI at 100 mg ml^{-1}. Store in the dark at 4°C.

Purification of DNA
7. To the DNA add 2 vol. 6 M NaI and mix.
8. Invert the silica suspension gently several times so that the particles are uniformly resuspended and add 0.5 µl silica suspension per 0.5 µg DNA. Leave for 5 min to allow the DNA to bind.
9. Spin the suspension 2000 g for 1 min.
10. Remove and discard the supernatant.
11. Wash the pellet by resuspending it in 0.5 ml 50 mM NaCl, 10 mM Tris-HCl, pH 7.5, 2.5 mM EDTA, 50% v/v ethanol. Centrifuge at 2000 g for 1 min.
12. Repeat step 11.
13. To elute the DNA, resuspend the pellet in 10–50 µl TE buffer, pH 8.0. Centrifuge at 2000 g for 1 min.
14. Remove the DNA-containing supernatant (note 1).

Note
1. Be careful not to take up any silica particles in the final DNA solution. Otherwise when the DNA is added to a high ionic strength solution, it will bind to the silica.

---

reporter molecules as up to about 30% of the probes may be extracted into the phenol phase.

The ease with which radioactivity can be detected makes it easy follow the separation of labeled probe and unincorporated nucleotides. Aliquots of fluorescein-labeled probes can be detected by spotting aliquots of eluates on to a filter and irradiating the filter with UV light. For biotin and DIG-labeled probes, aliquots of each fraction are spotted on to a filter and processed by the complete detection process – as described in the next section.

11.6 Determining the efficiency of incorporation of label

If labeling is inefficient or if too little probe is used, there may be so few reporter molecules present that only abundant targets can be detected after hybridization. This means that the sensitivity will be low. The presence of too many reporter molecules during hybridization is wasteful and may cause high backgrounds. So the efficiency of labeling is determined and the appropriate amount of labeled probe is subsequently used in hybridizations.

11.6.1 Radioactive probes

For radioactive probes, the specific activity is usually determined by measuring the radioactivity in an aliquot of the purified probe and comparing it to that in an aliquot of the total labeling mix in which the specific activity is known. Standard methods are used [1].

Cerenkov counting In Cerenkov counting, radioactivity is counted in a scintillation counter in the absence of added scintillant. This method can be used for strong emitters such as ^{32}P, but is unsuitable for weak emitters. The efficiency of counting ^{32}P in this way usually exceeds 90% and the method is very fast and the sample can be recovered and used after counting.

PROTOCOL 11.9: Cerenkov counting of ^{32}P-labeled probes

1. After the labeling reaction has been terminated, remove a 1–2 µl aliquot into a scintillation vial.
2. Similarly, place 1–5 µl of the purified probe in a vial.
3. Count by Cerenkov radiation, i.e. under ^{3}H settings without added scintillant.

TCA precipitation For all radiolabels, a sample can be removed and treated with ice-cold TCA which precipitates nucleic acid.

DE81 filter binding. DE81 filters (Whatman) are positively charged and bind nucleic acids and oligonucleotides strongly. The method is the same as that for TCA precipitation (*Protocol 11.10*) except that 0.5 M phosphate is substituted for TCA. This method is better for oligonucleotides <50 nts than TCA precipitation which is inefficient at precipitating oligonucleotides.

PROTOCOL 11.10: TCA precipitation

1. After the labeling reaction has been terminated, remove duplicate samples of 1–5 μl depending on the size of the labeling reaction and the amount of label added. Place on two Whatman GFC glass fiber filters. Label filters in pencil.
2. Set one filter aside to be counted with no further treatment. Place the other in about 20 ml ice-cold 5% TCA. Keep on ice with occasional shaking for 5 min.
3. Remove the filter and wash in a second aliquot of 5% TCA for 5 min.
4. Wash the filter in 70% ethanol for 5 min.
5. Air dry the filter.
6. Count both filters in a scintillation counter in the presence of scintillation fluid and under normal ^{32}P settings.

11.6.2 Calculation of the specific activity of a radioactive probe

The specific activity of the probe is defined as the c.p.m. incorporated μg^{-1} nucleic acid. A worked example is given for random priming in which there is net synthesis of DNA. To calculate the specific activity, it is first necessary to know how much DNA was synthesized and how many counts were incorporated.

Theoretical maximum yield if 100% label was incorporated =

$$\frac{Bq \ dNTP \ added \times 4 \times 330 \ ng \ nmol^{-1}}{specific \ activity \ dNTP \ (Bq \ nmol^{-1})}$$

The factor of 4 is required as there are four different nucleotides in the DNA; 330 is the average molecular weight of a dNMP.

The % incorporation of counts = $\dfrac{c.p.m \ incorporated \times 100}{total \ c.p.m \ added}$

Amount of DNA synthesized = $\dfrac{\% \ incorporation \times theoretical \ yield}{100}$

Specific activity = $\dfrac{Total \ c.p.m. \ incorporated \times 10^3}{(ng \ DNA \ synthesized + ng \ input \ DNA)}$

= c.p.m. μg^{-1}

The factor of 10^3 is used to change ng to μg.

Example. A sample of 1.85×10^6 Bq [α-^{32}P]dCTP (1.1×10^{14} Bq mmol^{-1}; 3.7×10^8 Bq ml^{-1}) is incubated in a standard 50 μl reaction containing 25 ng template DNA; 2 μl is removed, acid precipitated and the radioactivity counted.

Theoretical maximum yield $= \dfrac{1.85 \times 10^6 \text{ Bq} \times 4 \times 330 \text{ ng nmol}^{-1}}{1.1 \times 10^8 \text{ Bq/nmol}^{-1}}$

$= 22 \text{ ng}$

Radioactivity added to the reaction =

$1.85 \times 10^6 \times 60 \text{ c.p.m. Bq}^{-1} \times 90\%$ efficiency of counting (Cerenkov cts)
$= 9.9 \times 10^7 \text{ c.p.m.}$

Assume incorporation $= 2.51 \times 10^6 \text{ c.p.m. incorporated } 2 \, \mu l^{-1}$

Thus total incorporation $= \dfrac{2.51 \times 10^6 \times 55}{2}$

$= 6.9 \times 10^7 \text{ c.p.m. } 55 \, \mu l^{-1}$

(The aliquot was removed after addition of 5 μl stop mix, so total volume = 55 μl.)

% incorporation $= \dfrac{6.9 \times 10^7 \times 100}{9.9 \times 10^7}$

$= 70\%$

Amount DNA synthesized $= \dfrac{70 \times 22}{100}$

$= 15.4 \text{ ng}$

Specific activity $= \dfrac{6.9 \times 10^7 \times 10^3}{(25 + 15.4)}$

$= 1.71 \times 10^9 \text{ c.p.m. } \mu g^{-1}$

11.6.3 Nonradioactive probes

For nonradioactive labels, the efficiency of incorporation is estimated semiquantitatively by comparing the amount of label incorporated in the probe with standards containing known amounts of reporter molecule.

After the labeling is complete, an aliquot of the reaction mixture is removed and serial dilutions are made. Equal volumes of the dilutions are applied as dots to a nylon or charged nylon filter. The extent of dilution necessary will depend on whether the nucleic acid has been end or uniformly labeled and if there has been net synthesis of DNA as in random priming. Dots containing diluted standard are applied in parallel. The filter is washed to remove unincorporated nucleotides and the dots are visualized by the detection procedure appropriate to the label used.

In essence, this involves treating the filter with blocking agent, then with the Fab portion of an antibody which is conjugated to either horseradish peroxidase or alkaline phosphatase. The antibody fragment forms a link between the enzyme and the reporter molecule. Next, the bound enzyme reacts with added substrate to form a colored precipitate or to emit light depending on the substrate used. By comparing the intensity of the signal of probe dots with those of standards, the amount of incorporation can be estimated.

It is easy to make and store bulk preparations of labeled probe, so although determining the efficiency of labeling may seem tedious and inconvenient, it need not be carried out often.

The easiest way to monitor the efficiency of labeling of nonradioactive probes is to use a kit from a commercial company. The solutions have all been preoptimized and labeled standards are included against which the incorporation of label into the probe can be compared.

Fluorophores. An easy and rapid modification of the above method is available for determining the incorporation of nucleotides attached to fluorophores such as fluorescein. The label is visualized simply by irradiating with UV light which causes the fluorophore to glow brightly (*Protocol 11.11*).

An alternative approach for measuring the incorporation of non-radioactive label is to add a trace of radioactive NTP (or dNTP or ddNTP) such as 1.85×10^5 Bq [^{3}H]-dATP to the reaction mixture and to measure the efficiency of its incorporation. It is then assumed that the efficiency of incorporation of labeled and unlabeled nucleotides is the same.

PROTOCOL 11.11: Determining the efficiency of incorporation of fluorescein-labeled nucleotides

1. To a strip of nylon filter, apply 5 µl of serial dilutions (1/25, 1/50, 1/100, 1/250, 1/500, 1/1000) of the fluorescein-11-dUTP used to label the nucleic acid. Allow to dry in air. This is the control filter which is not processed further.
2. To a second strip, apply 5 µl of the reaction mix after the appropriate labeling time. Also apply a negative control such as 5 µl of a 1:5 dilution of the fluorescein-11-dUTP. Do not allow the spots on this second strip to dry before processing or it will be difficult to remove unincorporated nucleotide. Wash in about 100 ml 2 × SSC at 60°C with gentle shaking for 10 min.
3. Place both control and test strips face down on a transilluminator and view under UV light. There should be no color visible at the position of the negative control whereas the sample from the reaction mix should glow with a green/yellow color. A semiquantitative estimation of the incorporation level is made by comparing the intensity of the color with the dots on the control strip.
   A permanent record can be made by photography using an appropriate filter such as Kodak Wratten No. 9.

Note
1. The control strip can be stored at 4°C in the dark and re-used.

References

1. **Sambrook, J., Fritsch, E.F. and Maniatis, T.** (1989) *Molecular Cloning: a Laboratory Manual.* Cold Spring Harbor Laboratory Press, Cold Spring Harbor, New York.
2. **Hames, B.D. and Higgins, S.J.** (eds) (1995) *Gene Probes 1: A Practical Approach.* IRL Press, Oxford.
3. **Feinberg, A.P. and Vogelstein, B.** (1983) *Anal. Biochem.* **132:** 6–13.
4. **Feinberg, A.P. and Vogelstein, B.** (1984) *Anal. Biochem.* **137:** 266–267.
5. **Rigby, P.W.J., Dieckmann, M., Rhodes, C. and Berg, P.** (1977) *J. Mol. Biol.* **113:** 237–251.
6. **Langer, P.R., Waldrop, A.A. and Ward, D.C.** (1981) *Proc. Natl Acad. Sci. USA* **78:** 6633–6637.
7. **Sarker, G. and Sommer, S.S.** (1988) *Nucleic Acids Res.* **16:** 5197.
8. **Stoflet, E.S., Koeberl, D.D., Sarker, G. and Sommer, S.S.** (1988) *Science* **239:** 491–494.
9. **Kreig, P.A. and Melton, D.A.** (1987) *Methods in Enzymol.* **155:** 397–415.
10. **Kreig, P.A.** (1991) *Nucleic Acids Res.* **18:** 6463.
11. **Wahlberg, J., Hultman, T. and Uhlen, M.** (1995) In: *PCR2: A Practical Approach* (eds M.J. McPherson, B.D. Hames and G.R. Taylor). IRL Press, Oxford, pp. 71–86.
12. **Boyle, J.S. and Lew, A.M.** (1995) *Trends Genet.* **11:** 8.

12 Basic techniques: prehybridization, hybridization and washing

12.1 Equipment

Although sophisticated equipment is available for carrying out hybridizations, perfectly satisfactory results can be obtained with readily-available and inexpensive equipment.

One of the simplest methods of hybridization is to place the filter and appropriate solution in a polythene sleeve that is then heat-sealed and incubated in a water-bath. The sleeve must be opened and re-sealed when the probe is added and this carries the risk that traces of probe may contaminate the water-bath. This risk can be avoided by placing the sleeve in a plastic box containing water heated to the required temperature and which in turn is placed in a shaking water-bath. When handling many filters at once, it is more convenient to carry out prehybridization, hybridization and washing in plastic or polythene boxes rather than in sleeves. The boxes are incubated in a shaking water-bath.

Commercial devices are available which allow filters to be wound round an inner support that fits into a capped tube containing hybridization solution. The tube is then placed in a water-bath – not necessarily a shaking one. This procedure has the advantage that very little solution need be used which conserves probe and is very useful if reagents are limiting or expensive.

A hybridization oven can be used. The filter is placed round the inside wall of a hybridization bottle which fits into slots in the oven. The bottle rotates and continuously coats the surface of the filter with liquid.

Bottles of different dimensions are available to cater for different sizes of filter and volumes of solution.

12.2 Prehybridization

After nucleic acid has been firmly bound, the filter is ready for prehybridization. Some makes of filter wet unevenly if they are placed directly into prehybridization solution, probably because of the high ionic strength. This causes the prehybridization and hybridization solutions to coat the filter unevenly and gives rise to patchy hybridization. To avoid this happening filters can be pre-wet in a dilute detergent such as 1% Triton X-100 which wets the filter very evenly and quickly. Excess liquid is blotted off and the filter is then prehybridized.

Prehybridization serves two purposes: it equilibrates the filter with solution and it blocks sites at which the probe could bind non-specifically. Failure to prehybridize effectively can lead to patchy and unacceptably high backgrounds.

Prehybridization solutions typically contain a detergent, a blocking agent and a heterologous nucleic acid. The detergent is usually 1% SDS. It facilitates even covering of the filter with solution and has weak RNase-inhibiting properties. The most widely used blocking agent is Denhardt's solution (i.e. 0.02% Ficoll (mol. wt. 400 000), 0.02% polyvinyl pyrollidone (mol. wt. 400 000) and 0.02% bovine serum albumin [1]). Cheaper alternatives such as heparin [2] and reconstituted nonfat milk (which is also known as Bovine Lacto Transfer Technique Optimiser or BLOTTO) [3] can also be used.

The choice of blocking agent depends on the purpose of the hybridization. Denhardt's reagent tends to gives lower backgrounds than other reagents, so it is widely used when the copy number of sequences to be detected is low or the hybridization time is long [4]. Nonfat milk gives good results when screening recombinant phage and plasmid and also when probing Southern blots for abundant sequences. However, as components of milk tend to precipitate if SDS is present or if the incubation times are long, Denhardt's solution is preferred under these circumstances. Nonfat dried milk may contain variable amounts of ribonucleases and although they are reported to be completely inactivated by treatment with DEPC [5], it may be wise to avoid this blocking agent when using RNA on the filter or in the probe.

The nucleic acid in prehybridization solutions is usually calf thymus or salmon sperm DNA that has been sheared or sonicated to about 600 nt in length. It is denatured by heat treatment followed by snap cooling in ice to trap the DNA as single strands. To reduce backgrounds further, poly(A) is sometimes added. This is useful as a hybridization competitor if the probe or filter-bound sequences are rich in A or T residues, for example polyadenylated mRNA or cDNA made from it. Similarly, poly(C) can be included if the sequences are rich in G or C residues such as occur in GC-rich stretches in some genomes or when the recombinant being screened has been generated through oligo(dG) and oligo(dC) homopolymer tailing. For hybridizations involving RNA, yeast tRNA is often added to the prehybridization buffer in place of heterologous DNA.

The prehybridization solution is preheated to the desired temperature before adding to the filter. The choice of temperature will depend on whether aqueous or formamide-containing solutions are used and whether perfectly or imperfectly matched sequences are to be detected.

12.3 Hybridization

During hybridization, the labeled probe is incubated with the target sequences under conditions which allow the desired hybrids to form. The probe must be single-stranded. Double-stranded DNA probes are denatured by heat or alkali treatment. Heating can be either by placing the probe in a boiling-water bath or by treatment with microwaves [6]. If alkali treatment is used with a nonradioactively labeled probe, the spacer arm should not be linked to the nucleotide via an alkali-sensitive bond.

Several years ago it was customary to use prehybridization and hybridization solutions of different composition. The prehybridization solutions had higher concentrations of blocking agents and higher buffering capacity. Nowadays shorter prehybridization and hybridization times tend to be used and there has been a move towards using solutions of the same composition for prehybridization and hybridization.

The denatured probe may be added directly to the prehybridization solution covering the filter. This carries the risk of splashing undiluted probe on to the filter which creates very high backgrounds at the site of the splashes. It is also quite common, and safer, to use a fresh solution for hybridization. Changing to fresh solution minimizes pH change. It also has the additional advantage that a higher volume of solution can be used for prehybridization which ensures that the filter is well coated and equilibrated and allows the volume of hybridization solution to be low which is necessary to keep the concentration of probe high.

Solutions containing formamide are used for DNA:RNA hybridizations. In the past it was common to use formamide-containing solutions of different composition for DNA:DNA and DNA:RNA hybridizations. However, it is now current practice to use solutions with the same composition – apart from the formamide concentration itself.

For oligonucleotide hybridizations, solutions do not generally contain formamide, but hybridization in its presence is sometimes carried out [7].

A new family of hybridization solutions has been developed based on the presence of 7% SDS [8]. These solutions do not require blocking agents, carrier DNA or RNA or dextran sulfate. The basic solution contains SDS and a high concentration of NaCl, but substances such as formamide can be added to reduce the temperature of incubation and polyethylene glycol (PEG) to speed up the reaction [9,10]. The concentration of phosphate has to be increased in the presence of PEG in order to prevent the solution forming two phases [9]. Prehybridization is not required when positively charged filters are used as the presence of a high concentration of SDS keeps the background low.

The conditions for hybridization will depend on the purpose of the experiment and are chosen according to the guidelines in Chapters 7 and 8.

12.4 Washing

Filters are washed extensively after hybridization. This serves to remove both unreacted probe in solution and probe which binds loosely and nonspecifically to the filter. Washing also dissociates unwanted, mismatched hybrids. Washing solutions contain salt and a detergent, usually SDS, but do not contain formamide or other denaturing agents. Manipulation of the salt concentration and temperature of washing are important means of controlling stringency, which determines which hybrids are allowed to persist (see Chapters 4 and 5).

Filter-bound hybrids do not dissociate as quickly as hybrids in solution [8], so this must be taken into account when planning times of washing. With radioactive probes, it is easy to follow the efficiency of washing. Periodically a radiation monitor is held close to the filter and the radioactive signal can be seen to fall markedly as nonspecifically bound probe is removed. With experience, the signal that corresponds to an acceptable background can be determined.

It is not so simple to follow the efficiency of washing when nonradioactive probes are used. The filter is run through the detection procedure as described in Chapter 13 and if background is not low enough, the filter is re-washed. This process works when chemiluminescent substrates are used, but is not satisfactory when hybrids are visualized by precipitation of colored dyes. For the latter it is easiest to include a few replicate lanes or dots. After washing, a replicate lane is cut off the filter and developed while the main filter is kept in washing solution. If the background is too high, the main filter is washed again.

12.5 Protocols for prehybridization, hybridization and washing

There are many different protocols for filter hybridizations and no one will suit all applications. *Protocols 12.1–12.4* are designed as starting

---

PROTOCOL 12.1: Solutions for hybridization of DNA and RNA

Prehybridization/hybridization solution containing formamide		**Aqueous prehybridization/hybridization solution**	
Solution A		*Solution 1*	
• Deionized formamide (note 1)	50 ml		
• 20 × SSC	30 ml	• 20 × SSC	30 ml
• 100 × Denhardt's solution	5 ml	• 100 × Denhardt's solution	5 ml
• 20% SDS	0.5 ml	• 20% SDS	0.5 ml
• Water	14.5 ml	• Water	62.5 ml
Solution B		*Solution 2*	
• Sonicated DNA	2 ml	• Sonicated DNA	2 ml

Denature DNA by
either: placing in a boiling water bath 5 min
or: incubating in a microwave oven for
2 min at 600 W (note 6)
Chill in ice and add solution B to solution A.
Mix well.
Store at 4°C.

Denature DNA by
either: placing in a boiling water bath 5 min
or: incubating in a microwave oven for
2 min at 600 W (note 6)
Chill in ice and add solution 2 to solution 1.
Mix well.
Store at 4°C.

Notes
1. Formamide is a teratogen. Handle with care and use gloves.
2. For composition of stock solutions see Appendix D.
3. The composition of the formamide-containing prehybridization/hybridization solution is: 50% formamide, 6 × SSC, 5 × Denhardt's solution, 0.5% SDS, 100 μg denatured DNA ml^{-1}.
4. The composition of the aqueous prehybridization/hybridization solution is: 6 × SSC, 5 × Denhardt's solution, 0.5% SDS, 100 μg denatured DNA ml^{-1}.
5. For a hybridization solution containing dextran sulfate, omit water from solution A and add 10 g dextran sulfate (mol. wt. 500 000). Stir until dextran sulfate has dissolved. Adjust the volume to 98 ml and add solution B as above.
6. If the probe is dilute and the ionic strength is low, the microwaved DNA should not boil over. However, it is strongly recommended that practice runs are made and that times are modified if necessary to ensure that no boiling over occurs.
7. It is not usually necessary to filter prehybridization and hybridization solutions.

PROTOCOL 12.2: Hybridization of filters

Wetting the filter
1. Float the filter in a solution of 1% Triton X-100, taking care to prevent air bubbles being trapped underneath. When one side is wet, immerse the filter to wet the other side.
2. Remove the filter and blot gently on Whatman 3MM paper to remove excess liquid.

Prehybridization
3. Place the filter in a polythene sleeve that is sealed on three sides. Add the prehybridization solution (at least $0.2\,ml\,cm^{-2}$) prepared as in *Protocol 12.1* and pre-warmed to 42°C for the formamide-containing solution and 65°C for the aqueous solution. Gently squeeze out air bubbles and heat-seal the open end of the sleeve.
4. Place the sleeve in a box of water preheated to 42°C/65°C in a shaking water-bath. Set the bath to shake at a speed such that the liquid in the sleeves moves gently over the filter. Incubate for 2–16 h at 42°C/65°C as appropriate.

   Generally filters are left in prehybridization solution until just before adding the hybridization solution.

Hybridization
5. Denature the probe by placing in a boiling-water bath for 5 min. Snap chill in ice. See note 9 for determining the amount of probe to use.
6. *Either* cut the sleeve open on one side and add the probe quickly to the liquid. Take care that no probe splashes directly on to the filter. Gently squeeze out air bubbles and heat seal the open end of the sleeve. Ensure that the probe is well mixed into the hybridization solution.
   Or cut the sleeve open on one side and drain the liquid out. Roll a pipette over the surface of the bag to remove as much liquid as possible, but do not allow the filter to dry or high backgrounds will occur. Add the denatured probe to hybridization solution (at least $80\,\mu l\,cm^{-2}$) that has been prewarmed to 42°C/65°C. Mix quickly and add to the filter. Gently squeeze out air bubbles and heat-seal the open end of the sleeve.
7. Hybridize the filter at 42°C/65°C for the required time. Overnight is often convenient.

Washing
8. Cut open the bag and carefully decant the hybridization solution. Retain the solution if it is to be re-used, otherwise discard it in accordance with safety guidelines.
9. Remove the filter gently. Immerse it in 200 ml of 2 × SSC, 0.1% SDS at room temperature. Swirl the solution gently to wash probe off the filter. Rinse the filter twice for 5 min. each time in fresh aliquots of the same solution.
10. For a moderately stringent wash, wash the filter twice in 200 ml of 2 × SSC, 0.1% SDS at 60°C for 1 h each time. For a higher stringency wash, treat the filter 2 × 1h at 60°C, 0.1 × SSC, 0.1% SDS.
11. Finally, rinse the filter in 2 × SSC, at room temperature. Blot the filter to remove excess liquid, but do not dry if the filter is to be rewashed or rescreened.
12. Detect the hybrids as described in Chapter 13.

Notes
1. Several filters (10 at least) can be hybridized at once, but it is important that each filter is adequately covered with all solutions. If they stick together, patchy hybridization will occur as some filters will not have adequate exposure to the probe. In addition high backgrounds will occur because the probe has been trapped between filters.

   The secrets of successfully hybridizing many filters at once are first to place each filter individually into each solution, thus ensuring that they do not stick together and second to make sure that there is plenty of solution to cover the filters.

It is more convenient to carry out several hybridizations at once in a plastic box rather than a heat-sealed bag or hybridization tube. Wet each filter individually and place it individually in prewarmed prehybridization solution. After prehybridization, remove the filters and blot them lightly to remove excess liquid. Add denatured probe to prewarmed hybridization buffer in a fresh box and add the blotted filters individually making sure that they are all covered with solution. Work quickly so that the filters do not dry out. For washing, pour off the probe and place each filter individually in the washing solution. Periodically, change the position of the filters in the solutions so that they do not stick together.

2. It is **most** important that all prehybridization, hybridization and washing solutions are preheated to the appropriate temperature before they are added to the filter.

3. The formamide concentration and temperature of incubation can be altered according to the needs of the experiment, e.g. degree of mismatching that is to be permitted (see Chapter 7).

4. If applying probe directly to the prehybridization solution, add it into the solution and not to the sides of the sleeve or on to the filter. Mix well. It is generally safer for a beginner to replace the prehybridization solution with probe-containing hybridization solution as in the second option of step 6.

5. If hybridization accelerators are to be used, it is generally best to prehybridize in the absence of the accelerator and hybridize in its presence. This tends to give lower backgrounds. For the composition of hybridization solution containing dextran sulfate see note 4 of *Protocol 12.1*.

6. $6 \times$ SSPE should be used in place of $6 \times$ SSC for hybridizations with RNA probes. It can also be used in DNA:DNA hybridizations with long incubation times.

7. Denhardt's solution can be replaced by:
   either heparin at a concentration of $50 \, \mu g \, ml^{-1}$ in solutions lacking dextran sulfate or $500 \, \mu g \, ml^{-1}$ in hybridization solutions containing dextran sulfate.
   or 0.25% low-fat milk (original Marvel formulation and obtainable from grocery stores). Be wary of using low-fat milk for hybridizations involving RNA but see ref. [5].

8. For hybridizations involving RNA, yeast tRNA at $100 \, \mu g \, ml^{-1}$ and poly(A) at $10 \, \mu g \, ml^{-1}$ can replace sonicated DNA in both prehybridization and hybridization solutions.

9. Amount of probe to use. If the hybridization solution contains dextran sulfate as a rate enhancer, do not exceed a probe concentration of $10 \, ng \, ml^{-1}$ or high backgrounds may arise. If there is no dextran sulfate, the probe concentration can be increased to $50-100 \, ng \, ml^{-1}$. For a radioactive probe that has been labeled to a specific activity of $1-2 \times 10^9$ c.p.m. μg^{-1} a probe concentration of $10 \, ng \, ml^{-1}$ gives a solution of $1-2 \times 10^6$ c.p.m. ml^{-1}.

PROTOCOL 12.3: Hybridization with oligonucleotide probes

Prehybridization/hybridization solution

Solution 1
- $20 \times$ SSC   30 ml
- $100 \times$ Denhardt's solution   5 ml
- 20% SDS   0.5 ml
- Water   62.5 ml

Solution 2
- Sonicated DNA   2 ml

Denature DNA by *either*
- Placing in a boiling water bath 5 min

or
- Incubate in a microwave oven for 2 min at 600 W.

Chill in ice and add solution 2 to solution 1.

Mix well.
Store at 4°C.

Method

Wetting the filter
1. Wet the filter in 1% Triton X-100.
2. Remove the filter and blot gently on Whatman 3MM paper to remove excess liquid.

Prehybridization
3. Transfer the filter to the hybridization vessel (sac, tube or bottle) and add prehybridization solution prewarmed to the calculated incubation temperature. Add at least 0.2 ml cm^{-2} filter.
4. Incubate at the predetermined incubation temperature 1–4 h.

Hybridization
5. Add labeled oligonucleotide probe: 0.1–2 pmol ml^{-1} for tailed oligonucleotides or 1–10 pmol ml^{-1} if end-labeled. There is no need to denature it.
6. Incubate at predetermined temperature, 2–4 h.

Washing
7. Remove the filter and immerse it in 200 ml of 6 × SSC, at 4°C. Agitate the solution gently to wash the probe off the filter. Rinse the filter once more in 6 × SSC.
8. Wash filter three times for 15 min each time in 6 × SSC at the temperature determined as in Sections 8.5 and 8.6.
9. Finally, rinse the filter in 2 × SSC, at room temperature. Blot the filter to remove excess liquid, but do not dry if the filter is to be rewashed or rescreened.
10. Detect the hybrids as described in Chapter 13.

Notes
1. The composition of the hybridization solution is: 6 × SSC, 5 × Denhardt's solution, 1% SDS, 100 µg denatured DNA ml^{-1}.
2. The incubation temperature depends on the length and (G+C) content of the probe and whether a single or mixture of probes is being used (see Chapter 8).
3. Do not prolong the washing times as oligonucleotide hybrids dissociate very readily.
4. If pools of oligonucleotides are used, wash at a temperature of 5–10°C below the T_m of the oligonucleotide with the lowest T_m.

PROTOCOL 12.4: Hybridization in 7% SDS-containing solutions

Prehybridization/hybridization solution

20 × SSPE	4.5 ml
20% SDS	21 ml
Water	34.5 ml

Method
Wetting the filter
1. Wet the filter in 1% Triton X-100.
2. Remove the filter and blot gently on Whatman 3MM paper to remove excess liquid.

Prehybridization
3. Transfer the filter to the hybridization vessel, (sac, tube or bottle) and add prehybridization solution prewarmed to the calculated incubation temperature. Add at least 0.2 ml cm^{-2} filter.
4. Incubate at the predetermined incubation temperature for 1–4 h.

Hybridization

5. Denature double-stranded DNA probe by probe by placing in a boiling-water bath for 5 min. Snap chill in ice. If the probe is RNA or an oligonucleotide there is no need to denature it.
6. Remove the prehybridization solution.
7. Add the probe to prewarmed hybridization solution (see Note 2). Mix well and add to the filter. Incubate at the predetermined temperature for 2–4 h.

Washing

8. Remove the filter and immerse it in 200 ml of 6 × SSC, at room temperature for about 1 min. Agitate the solution gently to wash the probe off the filter. Rinse the filter once more in 6 × SSC.
9. Wash the filter according to steps 8 onwards of *Protocol 12.2* or *12.3* according to whether the probe is 'long' or an oligonucleotide.

Notes

1. The composition of the (pre-)hybridization solution is 1.5% SSPE, 7% SDS.
2. Use at least 0.125 ml hybridization solution per cm^2 filter.
3. Do not exceed a probe concentration of 50–100 ng ml^{-1} or high backgrounds may arise. For a radioactive probe that has been labeled to a specific activity of $1-2 \times 10^9$ c.p.m. μg^{-1} a probe concentration of 10 ng ml^{-1} gives a solution of $1-2 \times 10^6$ c.p.m. ml^{-1}.
4. Formamide can be included in the hybridization solution. Use the guidelines in Chapter 7 and adjust the temperature of incubation.

points which can be optimized as described in the guidelines in Chapters 7 and 8 according to the purpose of the experiment. *Protocol 12.2* can be used for both DNA and RNA hybridizations, *Protocol 12.3* is for oligonucleotide hybridizations.

It is most important that the filter is not allowed to dry out at any stage. If it dries before washing, probe will bind nonspecifically and irreversibly all over the filter causing high backgrounds. If the filter dries after washing, the probe will bind irreversibly at the site of hybrids and it may be impossible to strip it off for reprobing the filter.

It is also most important that all solutions are prewarmed to the appropriate temperature before being added to the filter. Otherwise the temperature that is eventually reached will depend on the length of the incubation. Failure to preheat solutions is a common source of variability in results.

Figure 12.1 shows the autoradiograph of a filter probed with an oligonucleotide from the J region of the human immunoglobulin l locus. The 20-mer probe was complementary to the mRNA and had the sequence: 5′-TTGG TCCC TCCG CCGA ACAC-3′. The composition was G4: C9: A3: T4, and calculated $T_m = 4(G+C) + 2 (A+T) = 52+14 = 66°C$. The temperature of prehybridization and hybridization was 42°C. This is permissive (i.e. allows for detection of mismatches) because the exact sequence of the target was unknown and variability in J_λ sequences is common. Washing was as in *Protocol 12.3* with step 8 at 4°C and step 9 at room temperature.

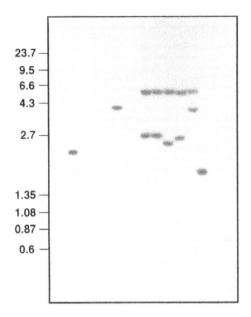

Figure 12.1. Hybridization with an oligonucleotide probe. A blot containing restriction digests of a series of recombinant plasmids carrying human IgC$_\lambda$ genes was probed with a ^{32}P-labeled antisense J$_\lambda$ oligonuclotide probe. The sequence of the oligomer was 5'-TTGGTCCCTCCGCCGAACAC-3'. Hybridization was in 6 × NET, 5 × Denhardt's solution, 0.1% SDS, 50 µg calf thymus DNA/ml for 13 h, 42°C. Washing was twice in 6 × SSC at 4°C for 15 min each time, then three times in 6 × SSC at room temperature for 15 min each time.

References

1. **Denhardt, D.T.** (1966) *Biochem. Biophys. Res. Commun.* **23:** 641–645.
2. **Singh, L. and Jones, K.W.** (1984) *Nucleic Acids Res.* **12:** 5627–5638.
3. **Johnson, D.A., Gautsch, J.W., Sportsman, J.R. and Elder, J.H.** (1984) *Genet. Anal. Tech.* **1:** 3–8.
4. **Sambrook, J., Fritsch, E.F. and Maniatis, T.** (1989) *Molecular Cloning: a Laboratory Manual.* Cold Spring Harbor Laboratory Press, Cold Spring Harbor, New York.
5. **Siegel, L.I. and Bresnick, E.** (1986) *Anal. Biochem.* **159:** 82–87.
6. **Stroop, W.G. and Schaefer, D.C** (1989) *Anal. Biochem.* **182:** 222–225.
7. **Albretsen, C., Haukanes, B.-I., Aasland, R and Kleppe, K.** (1988) *Anal. Biochem.* **170:** 193–202.
8. **Church, G.M. and Gilbert, W.** (1984) *Proc. Natl Acad. Sci. USA* **81:** 1991–1995.
9. **Amasino, R.M.** (1986) *Anal. Biochem.* **152:** 304–307.
10. **Budowle, B. and Baechtel, F.S.** (1990) *Appl. Theor. Electro.* **1:** 181–187.

13 Basic techniques: detection of hybrids

13.1 Radioactive hybrids

Hybrids containing radioactive isotopes are detected by virtue of the energy emitted by the isotope. Autoradiography is the most commonly used technique for detecting ^{32}P-labeled hybrids.

13.1.1 Autoradiography

In autoradiography, a filter containing radioactive hybrids is placed in close contact with an X-ray film in a light-proof cassette. Decay of isotope results in the emission of β-particles which react with silver halide grains on the film to create a stable latent image. The latent image gives rise to a visible image when the film is developed. The position of the image coincides with the position of hybrid on the filter and the intensity of the image is a measure of the amount of hybrid present.

X-ray film generally consists of a flexible plastic sheet coated on both sides with a photographic emulsion and covered by an outer layer of material such as gelatine which serves to protect the emulsion from scratches. The film can readily capture images formed through decay of medium strength β-emitters such as ^{35}S and ^{14}C. However, with strong β-emitters such as ^{32}P most of the energy passes straight through the film without interacting with the emulsion (*Figure 13.1*). This causes the sensitivity to be rather low.

Indirect autoradiography. To improve detection of ^{32}P, the X-ray film is sandwiched between an intensifying screen and the filter in the cassette. Intensifying screens are flexible plastic sheets coated with a scintillator such as calcium tungstate which emits light when activated. The energy emitted by ^{32}P passes through the film and strikes the

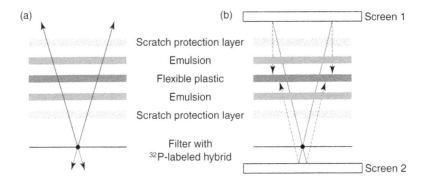

Figure 13.1. Autoradiography of ^{32}P-containing hybrids. (a) Direct autoradiography. The filter is placed in contact with the X-ray film. Radiation emitted by the ^{32}P on the filter reaches the film, but most passes through without exposing it. (b) Indirect autoradiography. Radiation emitted by the ^{32}P passes through the film, but hits an intensifying screen. Light is reflected back on to the film exposing it.

intensifying screen where it is converted into light which is reflected back on to the film creating a latent image (*Figure 13.1*). This procedure increases detection of ^{32}P about 10-fold. If a second screen is present on the other side of the filter, a further twofold enhancement is obtained. The latent image created by reflection of light is stabilized at low temperatures, so it is customary to place the cassette in a freezer at $-70°$C when using intensifying screens.

Although sensitivity of detection of strong β-emitters is increased with intensifying screens this is at the expense of a degree of resolution, i.e. the image is less sharp. However, loss of some resolution is usually acceptable in order to increase the sensitivity and enable rare sequences to be detected.

When the radiation energy is converted into light, the response of the autoradiographic film to the signal is not linear and in particular low amounts of ^{32}P are detected inefficiently. Film can be made to have a linear response by exposing it to an instantaneous flash (about 1 ms) of light then placing it with the flashed side against the intensifying screen and the filter on the other side. This procedure, termed preflashing, increases the sensitivity of detection, but it can be difficult to obtain consistent results. Preflashing units are available commercially and instructions that accompany them should be followed closely.

Choice of films. A wide range of different X-ray films is available for autoradiography. The choice of film should take into account:

- the isotope being detected;

- the sensitivity and resolution required;
- whether direct or indirect autoradiography is to be used;
- the color of light emitted by intensifying screens;
- the method used to develop the autoradiograph.

Film designed for high-energy β-emitters should be used for detection of ^{32}P. Those for weaker emitters, such as ^{14}C, ^{35}S and ^{3}H, are generally coated on only one side with photographic emulsion and for ^{3}H lack a protective antiscratch layer to enable the isotope to be in very close contact with the emulsion.

Film with emulsion on one side gives very high resolution and is commonly used in direct autoradiography. Film designed for direct autoradiography tends to be insensitive to blue light. Film with emulsion on both sides is best for autoradiography with intensifying screens as the sensitivity is greater than with single-coated films. The film should be sensitive to the color of light emitted by the scintillator in the intensifying screen. For example, calcium tungstate emits blue light and so film with optimal sensitivity at this part of the spectrum should be used.

Some films are designed for developing in automatic processors whereas others are designed for manual development in tanks. The type of film used should be appropriate for the developing system that is available.

Setting up the autoradiograph. If the filter is to be rewashed or reprobed, it must not be allowed to dry out or the probe will bind irreversibly. However, a wet filter will stick to the film and cause high background readings and may even tear when being separated from the film. To prevent this happening a thin layer of material is placed between the film and the filter. After the last wash, the wet filter is blotted with a tissue to remove most of the moisture and placed into thin polythene sleeving or covered in Saranwrap or Cling film before being placed in contact with the X-ray film. The filter must not be too wet otherwise ice crystals will form when the cassette is placed at $-70°C$. This may distort the filter and could cause it to crack.

It is useful to cut off the top left hand corner of the film. This is aligned with the top left hand corner of the filter and allows for easy orientation. The autoradiograph can be annotated with alignment marks and information such as the date and details of the experiment. A simple and inexpensive method is to make some radioactive ink by adding any ^{32}P-labeled compound to ordinary waterproof ink. Alignment marks etc. are written in the radioactive ink on adhesive tape which is then affixed to the edge of the filter before setting up the autoradiograph. The amount of radioactivity to use has to be determined empirically, but as a start point a solution can be prepared containing sufficient ^{32}P to show 500–2000 c.p.s. on a hand-held minimonitor [1]. This can be tested on labels exposed to film for different lengths of time and diluted if necessary. A more expensive option is to use commercially available pens which mark in radioactive ink. Nonradioactive options are luminescent peel-off stickers and pens which mark in phosphorescent inks.

The autoradiograph should be set up in a dark room with a red safelight (*Protocol 13.1*).

PROTOCOL 13.1: Setting up and developing an autoradiograph

Setting up an autoradiograph
For direct autoradiography
1. Place the filter in a light-proof cassette and place an X-ray film on top.
2. Close the cassette and leave it at room temperature to expose.

For indirect autoradiography
1a. Wipe the intensifying screen(s) with antistatic solution.
1b. *Either* place the filter in the cassette, place the X-ray film on top and finally place the intensifying screen on top so that the surface coated with scintillator faces the film; *or* if two intensifying screens are used, place one below the filter and the other on top of the film with the surfaces coated with scintillator facing each other.
2. Close the cassette and place it in a freezer at $-70°$C.

Development of film
Direct autoradiography
3. In the dark room and using a red safety light, open the cassette and place the film in an automatic X-ray processor or develop by hand. The process of development varies according to the type of film used, so manufacturer's recommendations should be closely followed. Care needs to be taken with manual development to prevent the film being scratched.

Indirect autoradiography
3. In the dark room and using a red safety light, remove the cassette from the freezer and allow it to come to room temperature before developing. If the film is cold when it is processed, condensation will occur which will fog the film.

Notes
1. To prevent static electricity do not wear gloves when handling film. Handle the film only at the edges.
2. The time of exposure of filter to the film varies according to the level of radioactivity in the hybrids and ranges from several hours to about 2 weeks for ^{32}P. There is no advantage in exposing the film for longer than the half-life of 14 days. With experience the time of exposure can be estimated by scanning the filter after the last wash with a hand-held monitor. If there is a slow 'click-click' sound, the sample may need exposure of several days to a week. The faster the clicks follow each other, the shorter the exposure time needs to be. Since autoradiography is a nondestructive technique, different exposure times can be tried with the same filter. As little as 5–10 c.p.m. above background can be detected by autoradiography using intensifying screens.
3. Images on an autoradiogram may have very different intensities, so when the filter has been exposed to film for a time that gives an acceptable intensity for some bands, other bands may be barely visible. Or, if the autoradiogram is exposed long enough to visualize the faintest images, the most intense bands may be overexposed, black and

very 'fuzzy'. One solution is to expose the film for different times as discussed in note 2 above. Alternatively, if preflashing facilities are available, it may be possible to get an acceptable level of signal from all bands on a single exposure by extending the linear range by preflashing.

13.1.2 Phosphor imaging

Radioactive hybrids may be detected by phosphor imaging. In this procedure radiation emitted by the isotope hits a special phosphor imaging screen creating a latent image. The image is scanned and a computer-generated image produced. The length of time required to obtain an image depends on the amount of radioactivity present. It is claimed that phosphor imaging is between 10 and 100 times more sensitive than autoradiography. The resolution is not yet as good, but the linear range of detection is superior covering six orders of magnitude compared with two to three for autoradiography. Screens can be reused after they have been bleached by exposure to light.

The main disadvantages of phosphor imaging are as follows.

- The machines are very costly and beyond the means of many laboratories.
- The degree of resolution is less than with autoradiography.

13.2 Detection of nonradioactive hybrids

Methods for detecting hybrids containing nonradioactive hybrids all depend on enzymatic activity. The enzymes most commonly used are alkaline phosphatase (AP) and HRP but β-galactosidase can also be used.

Direct detection. If the enzymes themselves are the primary labels for the probe, they are already present at the site of hybrids and detection can start as soon as the filter has been washed. Detection simply involves reacting the enzymes with chromogenic or chemiluminescent substrates so that a colored precipitate or flash of light is generated. The precipitate or flash of light is localized to the site at which the enzyme and consequently the hybrid is present. The colored precipitate is visible by eye whereas light can be detected on X-ray film that is sensitive to blue light.

Detection is straightforward and fairly sensitive. The disadvantage of direct detection is that the conditions of hybridization and washing must be very gentle in order to protect enzyme activity. This limits the stringency with which hybridization can be carried out. So direct detection is most often carried out with oligonucleotide probes which are normally hybridized at low temperature or with larger probes, but in the presence of formamide or urea.

Indirect detection. Detection of hybrids containing digoxygenin, fluorescein and biotin requires an extra step in order to direct the enzyme to the site of hybrids. The reporter molecules commonly used in filter hybridization are all highly immunogenic so it is common to use label-specific antibodies (Ab) conjugated to either AP or HRP to link the hybrid and enzyme (*Figure 13.2*). Detection involves sequential steps of adding antibodies conjugated to one of the enzymes and then supplying appropriate substrates.

Biotin in hybrids can also be linked to enzymes via streptavidin, a protein isolated from *Streptomyces avindii*. which has very high affinity for biotin. Streptavidin carries four deep-seated sites at which biotin binds. (The biotinylated nucleotides used to label the probe have long spacer arms so that biotin can reach the binding sites in streptavidin without being impeded by nucleic acid.) Several streptavidin-based detection methods are available. One of the simplest is to add streptavidin coupled to either AP or HRP to form a bridge between the biotinylated hybrids and the enzyme. The appropriate substrate is then added (*Figure 13.3*).

Avidin is a basic glycoprotein, isolated from egg white, that binds biotin with as high an affinity as streptavidin. However, avidin tends to bind nonspecifically to proteins particularly those containing lipid. This can give rise to unacceptably high backgrounds. So streptavidin is preferred to avidin for biotin detection.

Indirect detection tends to be preferred to direct methods because higher stringency of hybridization can be used.

13.2.1 *Chromogenic substrates*

The most sensitive chromogenic substrate system for alkaline phosphatase is 5-bromo-4-chloro-3-indolyl-phosphate/nitroblue tetrazolium (BCIP/NTB) (*Figure 13.4a*). Color formation is initiated when alkaline phosphatase catalyzes the removal of a phosphate from BCIP. A color-producing cascade occurs that finally gives a precipitate of diformazan which appears brown on nylon filters and blue on nitrocellulose filters [2]. The intensity of color reflects the number of enzyme molecules reacting.

An alternative and cheaper detection system is based on the use of different 2-hydroxy-3-naphthoic acid anilide (naphthol AS) phosphate substrates in the presence of different diazonium salts. [3]. Brightly colored

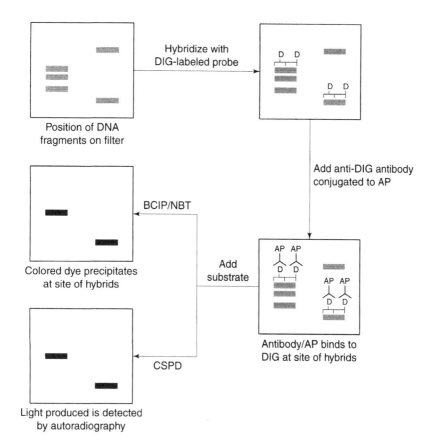

Figure 13.2. Detection of nonradioactive hybrids containing DIG. Hybrids containing DIG react with antibody against DIG. The antibody is conjugated to alkaline phosphatase which reacts with the appropriate substrate to produce a colored precipitate or flash of light at the site of the hybrids. Similarly, hybrids can also be detected using antibody conjugated to HRP and appropriate substrates.

azo dyes are produced with the color depending on which substrate/salt combination is used. The background, speed of development and resolution are superior to BCIP/NBT, but the sensitivity is less.

In the presence of HRP and hydrogen peroxide, the chromogenic substrates 4-chloro-1-naphthol gives rise to a bright purple precipitate. The sensitivity of detection is not very good although it can be improved by viewing under UV light [4]. An alternative chromogen is 3,3',5,5'-tetramethylbenzidine which in the presence of dextran sulphate yields a blue product at the site of hybrids [5].

Multiprobe detection. A useful feature of chromogenic detection is that more than one hybrid can be detected at once [6]. Each probe is

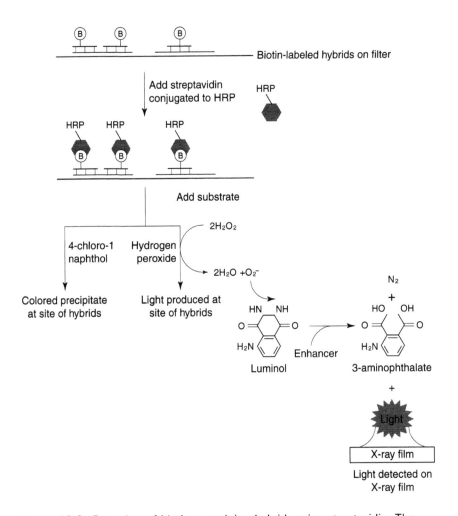

Figure 13.3. Detection of biotin-containing hybrids using streptavidin. The biotinylated probe hybridizes to target nucleic acid. Streptavidin coupled to HRP binds to biotin. On addition of luminol and an enhancer, light is produced which is detected by autoradiography. Detection can also be carried out by dye precipitation if chromogenic substrates are used. Similarly, detection can be carried out using AP conjugated to streptavidin together with an appropriate substrate.

labeled with a different reporter molecule, for example biotin, fluorescein and digoxygenin and they are simultaneously hybridized to the filter-bound nucleic acid. The different hybrids are visualized sequentially using different combinations of streptavidin– or antibody–enzymes and substrates (e.g. streptavidin–alkaline phosphatase, anti-fluorescein Ab–alkaline phosphatase and anti-DIG-Ab–alkaline phosphatase and three different naphthol-AS-phosphate/diazonium salts as substrates) (*Figure 13.5*). These yield different colors according to the substrate/salt combination. Between each detection reaction, the filter is treated with EDTA at high temperature (about 85°C) to inactivate the previous enzyme. After the final reaction, the filter shows red, blue and green signals corresponding to the sites of hybridization of the three probes. If the target sequence hybridizes to more than one of the probes, a mixed color is obtained at that site.

Precautions need to be taken to ensure that the high-temperature treatment used to inactivate alkaline phosphatase in multiprobe detection does not dissociate the hybrids. So, after washing the filter, but before detection, the hybrids are fixed to the filter by UV cross-linking. This precludes the use of nitrocellulose filters because of the risk of fire. In multicolor detection about 0.3 pg DNA can be detected in 2 h.

There are several disadvantages of chromogenic detection systems. The main one is that they lack the sensitivity of radioactive methods. Covalently bound probes can not be removed from filters after multiprobe detection and it may be difficult to remove the colored precipitate prior to reprobing the filter. This may not matter if the next probe is detected by a different means such as chemiluminescence.

13.2.2 Chemiluminescent substrates

The development of chemiluminescent substrates for AP and HRP has significantly improved the sensitivity with which hybrids can be detected. Chemiluminesence occurs when the energy released in a reaction is emitted as light.

Stable, substituted dioxetane substrates have been developed that cause light to be emitted when they are hydrolyzed by alkaline phosphatase. The most commonly used substrates for hybrid detection are sold under the trade names of AMPPD® and CSPD® and CDP-Star (*Figure 13.4b*). The compounds are very stable until the phosphate group is cleaved. This leads to formation of a metastable intermediate which decomposes and emits light at 477 nm. The light signal can be captured on X-ray film or by phosphor imaging using specialized photoexcitable storage phosphor screens [7]. The intensity of light emitted is a measure of the amount of enzyme activity which in turn is a measure of the amount of hybrid. The time taken for steady-state emission of light and the duration of light production varies according to the substrate.

The intensity of light emission can be increased by inclusion of the dioxetane substrate in micelles (Lumigen) or in covalently linked

(a)

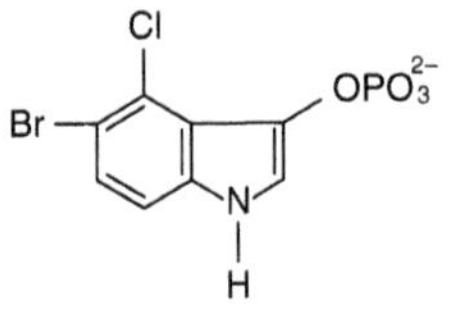

BCIP

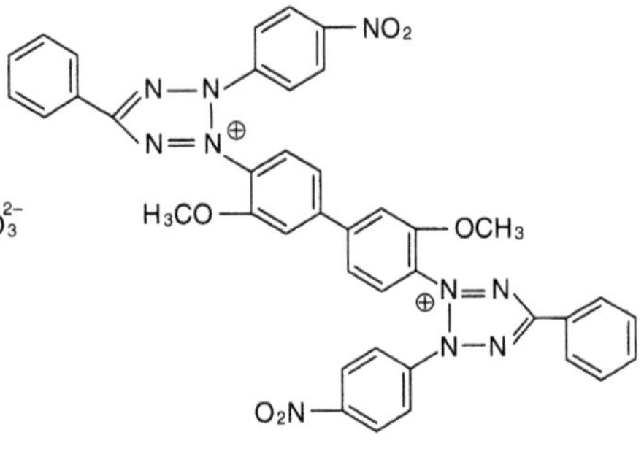

NBT

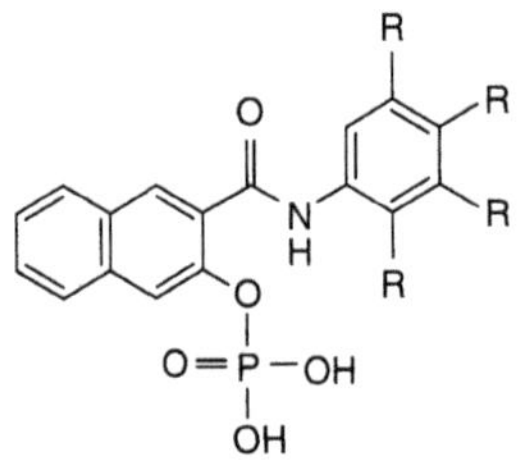

Naphthol AS phosphate
(R = −Cl or −CH₃)

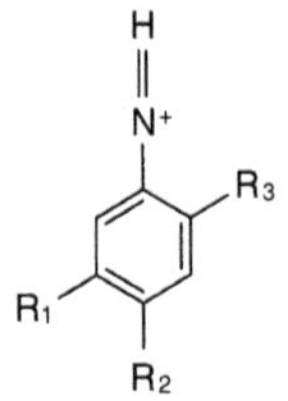

Diazonium salt

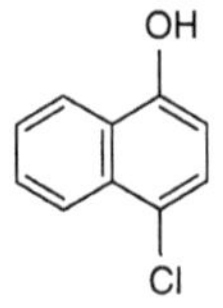

4-chloro-1-naphthol

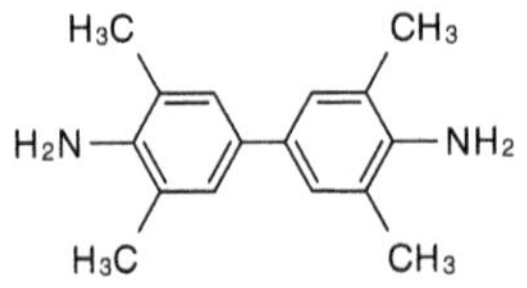

3,3', 5,5'-tetramethylbenzidine

(b)

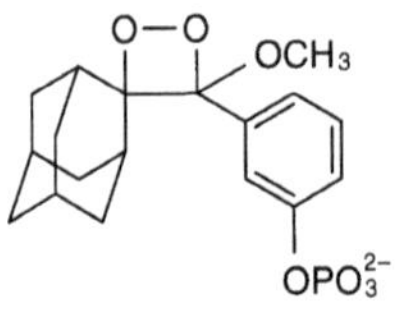

AMPPD®

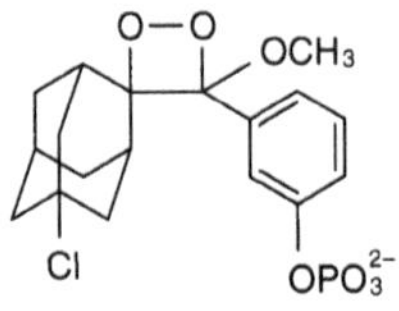

CSPD®

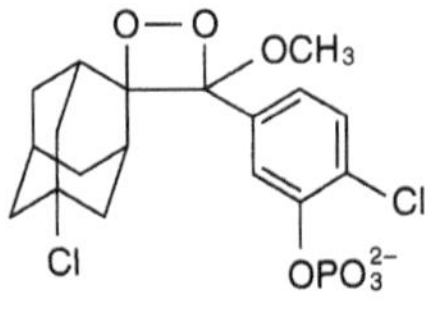

CDP-Star

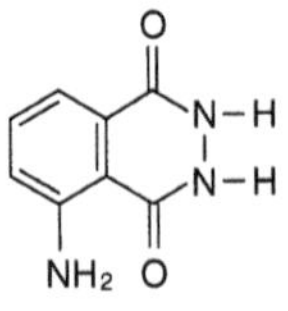

Luminol

Figure 13.4. Commonly used substrates for detection of nonradioactively labeled probes. (a) Chromogenic substrates BCIP/NBT and naphthol AS phosphate/ diazonium salts are used for detection of alkaline phosphatase bound to the probe. R = -Cl or -CH$_3$. R1, R2 and R3 differ according to the identity of the diazonium salt. 4-chloro-1-napthol and 3,3', 5,5'-tetramethylbenzidine are substrates for horseradish peroxidase. (b) Chemiluminescent substrates AMPPD, CSPD and CDP-Star are dioxetane substrates for alkaline phosphatase: Luminol is a substrate for horseradish peroxidase.

---

polymers which also contain fluorophores that are activated by the emitted light. For example, Lumiphos™ 530 contains cetyltrimethylammonium bromide and fluorescein surfactant (Boehringer Mannheim).

The AMPPD is rather hydrophobic, but the chloride groups in the other dioxetane substrates improve their solubility in water. These dioxetanes can be used for the same applications, but the kinetics of light emission and the time for which light emission occurs differs with each. CDP-Star is one of the newest and most sensitive dioxetane substrates so far developed. Exposure times of a few minutes are sufficient to detect a single copy gene in 1 μg of DNA on a Southern blot. Because light emission lasts for many hours, multiple exposures can be taken.

Chemiluminesence detection can be performed with both nylon and nitrocellulose filters, but longer exposure times are required for nitrocellulose and it may be necessary to use blocking agents (e.g. Nitroblock® from Tropix Inc.) for similar signal intensity.

13.2.3 Enhanced chemiluminescence

Horseradish peroxidase catalyzes the oxidation of luminol to 3-amino-phthalate via several intermediates. The reaction is accompanied by emission of low intensity light at 428 nm. However, in the presence of certain chemicals, the light emitted is enhanced up to 1000-fold making the light easier to detect and increasing the sensitivity of the reaction [8]. The enhancement of light emission is called enhanced chemiluminescence (ECL). Several enhancers can be used, but the most effective are modified phenols, especially p-iodophenol. The intensity of light is a measure of the number of enzyme molecules reacting and thus of the amount of hybrid [9].

ECL is simple to set up and is sensitive, detecting about 0.5 pg nucleic acid in Southern blots and in Northern blots [10]. Commercial detection kits are available that have been preoptimized and are easy to use.

The detection of nonradioactive hybrids is usually carried out with commercially available kits. The steps involved in detection are shown

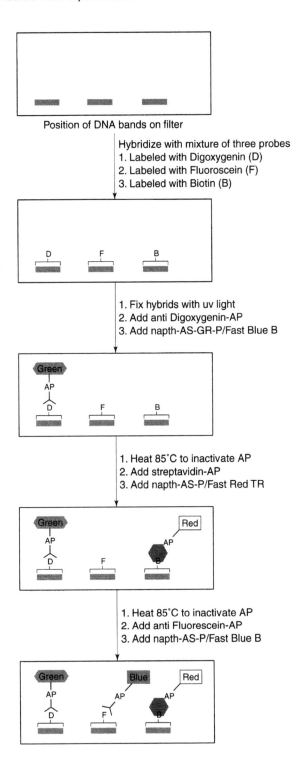

Position of DNA bands on filter

Hybridize with mixture of three probes
1. Labeled with Digoxygenin (D)
2. Labeled with Fluoroscein (F)
3. Labeled with Biotin (B)

1. Fix hybrids with uv light
2. Add anti Digoxygenin-AP
3. Add napth-AS-GR-P/Fast Blue B

1. Heat 85°C to inactivate AP
2. Add streptavidin-AP
3. Add napth-AS-P/Fast Red TR

1. Heat 85°C to inactivate AP
2. Add anti Fluorescein-AP
3. Add napth-AS-P/Fast Blue B

Figure 13.5. Multiple probe detection allows more than one sequence to be detected at a time. A blot is hybridized with several probes simultaneously, each probe being labeled with a different reporter group. The probes are detected sequentially using AP and a different combination of chromogenic substrates for each probe.

below, but the buffers, dilutions of antibodies, volumes, times and temperatures of incubation are specific to the kit used and should be followed closely.

Advantages of chemiluminescent detection. Detection by chemiluminescent substrates has several advantages over chromogenic substrates. The sensitivity is some 10–100-fold greater and quantitation of light emission is possible over a wide dynamic range whereas that for colored precipitates is much more limited, possibly only over one order of magnitude. Stripping filters is much easier when chemiluminescent substrates are used. Colored precipitates can be difficult to remove efficiently and the probes can not be removed by multiple probe detection because they will have been fixed by UV treatment.

Figure 13.6 compares the detection of RNA by dye/precipitation and ECL.

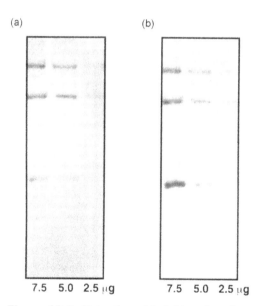

(a)   (b)

7.5   5.0   2.5 μg   7.5   5.0   2.5 μg

Figure 13.6. Detection of hybrids using chromogenic and ECL substrates. Detection of wound-induced genes in genomic DNA extracted from wounded leaves of hybrid poplar H11-11. Eco RI-digested poplar DNA was probed with win8 DNA, DIG-labeled by PCR. Hybridization was with high concentration SDS buffer + 50% formamide, 37°C overnight. Stringent washes were used. (a) Color detection with NBT/BCIP. (b) Chemiluminescent detection with AMPPD. Exposure time 60 min.

PROTOCOL 13.2: Detection of nonradioactive labels in hybrids (based on the use of commercial kits)

1. After the final posthybridization wash, blot the filter lightly, but do not allow it to dry.
2. Equilibrate the filter in wetting solution that usually contains a buffer, antibodies and a detergent.
3. Add a solution of blocking agent (often bovine serum albumen) to prevent enzyme substrates from binding to the filter and causing high backgrounds. Treat at the concentration and for the time recommended by the manufacturer.
4. Dilute the antibody (against the reporter molecule) by an appropriate factor in blocking buffer. Remove the liquid from the filter and add diluted antibody. Incubate for the recommended time.
5. Discard the antibody solution and wash several times in washing buffer to remove excess unbound antibody.
6. Equilibrate the filter in detection buffer.

For chromogenic detection
7. Prepare the substrate solution, e.g. BCIP/ NBT, in detection buffer.
8. Remove the detection buffer from the filter and replace it with substrate solution. Keep the filter in the dark and do not shake it.
9. Color development starts within a few minutes and can proceed for 12 h or more. The progress of color development can be followed by exposing the filter to light for a short time. Do not overdevelop.
10. Stop color development by washing the filter in an appropriate solution.
11. Dry the filter in air unless it is to be reprobed when it can be stored, sealed in a polythene sleeve at 4°C.
12. The results can be recorded by photography or by photocopying on to transparent sheets.
13. The color on the filter will fade with time. It may be possible to restore it by wetting the filter in a suitable solution (e.g. TE buffer for diformazan precipitates) provided that it has not been exposed to strong light.

For chemiluminescence or ECL detection
7. Prepare the substrate solution, e.g. luminol + hydrogen peroxide + *p*-iodophenol (enhancer) in detection buffer.
8. Remove the detection buffer from the filter, but do not allow it to become even slightly dry or high backgrounds will occur. The following step is designed to add just enough substrate solution to the filter without wasting expensive chemicals.
Either place the filter between clear photocopying sheets or in a clear pocket of the type used to file sheets of paper and add just enough substrate solution to cover the filter; *or* dip the filter in substrate solution and transfer it to a polythene sleeve and heat seal.
9. Transfer the filter to an X-ray cassette and expose to a film that is sensitive to blue light. Special photographic films and light-sensitive papers have been developed for chemiluminescence detection.

The amount of time that elapses until steady-state emission of light is reached will depend on the substrate/enzyme system used. Follow the manufacturer's recommendations. Multiple exposures can be taken for up to about 2 days after substrate is added (note 1).

Alternative streptavidin-binding procedure for detecting biotin
Carry out steps 1–3 as above
4a. Dilute streptavidin in wetting solution. Incubate filter in dilute streptavidin solution to allow streptavidin to bind to the biotin label in the hybrids.
4b. Wash the filter several times in wash solution to remove unbound streptavidin.
5a. Add a solution of enzyme conjugated to biotin. The enzyme will bind to the streptavidin by virtue of the biotin to which it is conjugated.

5b. Wash the filter in washing solution to remove free enzyme.

Carry out steps 6 onwards as above according to the enzyme used and whether chromogenic or chemiluminescent detection is to be used.

Notes

1. Detection of chemiluminescence: The time for which film should be exposed to autoradiographic film or exposure for (chemi)luminescence depends on the number of reporter molecules in the hybrids and the substrate used for detection. Because of the kinetics of light emission, exposures for CSPD are longer than those for CPD-Star. Typical exposures are 15–30 min and less than 1 min, respectively. If using a kit, follow the manufactureris advice.

Quantitation of hybridization signals. For many applications of filter hybridization, it is not necessary to quantitate the hybridization signal. Visual inspection of the autoradiograph or the colored precipitate on a filter can give all the information that is needed. For example, in colony/plaque hybridizations the presence and location of a hybridization signal may be all that is required. Differences in intensity of the signal can often be assessed by eye and no further quantitation may be needed.

In semiquantitative studies more accurate information is required and the images on an autoradiograph can be scanned electronically. This is easy and reliable. With a basic scanner the bands are scanned and the area under peaks integrated. With appropriate standards a calibration curve can be constructed with the area under the peaks being plotted against known amounts of nucleic acid (*Figure 13.7*) [11,12]. The intensity of the signal of an unknown sample, can be compared with the intensity generated by the standards. With computer controlled scanners the information from the scan is stored in the computer memory and there are more options for analysis. For example, areas of interest can be magnified and highlighted by altering the contrast and removing background. The results can be fed into other computer programs such as databases or spreadsheets.

The autoradiograph must not be overexposed. A problem with quantitating the hybridization signal via autoradiography is that X-ray film has such a low dynamic range, i.e. the intensity of signal is proportional to the amount of nucleic acid only over a small range of amount of nucleic acid. One way to overcome this is to cut ^{32}P-labeled spots/bands of the standards and unknown from the filter and count the radioactivity in a scintillation counter.

13.3 Reuse of filters and probes

Reprobing the same filter with a series of different probes can yield valuable information. It is particularly useful if the source of material on

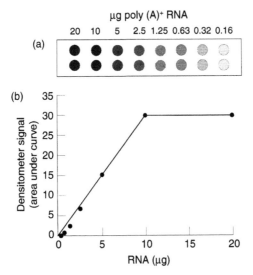

Figure 13.7. Quantitation of hybridization signals by densitometry. (a) A dot blot carrying standard amounts of nucleic acid is hybridized with a probe. After visualization of the hybrids each line of dots is scanned with a densitometer to measure the intensity (darkness) of the hybridization signals. (b) For each dot, the area under the curve is integrated and plotted against the amount of nucleic acid complementary to the probe in an unknown sample can be determined by hybridizing and scanning and comparing the density of the signal with the calibration curve.

the filter is scarce. Before probing the filter with a new probe it is first necessary to strip off the previous probe and to remove precipitates produced by nonradioactive detection systems. Removal of precipitates and stripping should be undertaken as soon as a satisfactory hybridization signal has been obtained.

13.3.1 *Removal of products of nonradioactive detection systems*

The substrates and products of nonradioactive detection systems are usually removed before the probe is stripped off. Luminescent or chemiluminescent products can generally be removed by a short wash in water. The purple diformazan precipitate produced by the BCIP/NBT staining method can be removed by dimethylformamide (DMF) treatment at high temperature. Since DMF dissolves nitrocellulose, this type of filter should not be used if reprobing is required. DMF is flammable and volatile and unless the material on the filter is precious, it is better to avoid removal of the precipitate altogether and to prepare a new filter. A protocol for DMF treatment is not given, but reference [9] can be consulted for details.

Naphthol AS phosphate/diazonium salt precipitates can be removed by ethanol treatment (*Protocol 13.3*).

PROTOCOL 13.3: Removal of substrates and products of nonradioactive detection systems

Chemiluminescent substrates
1. Wash the filter in water for 1–5 min.
2. Proceed to removal of probe – *Protocol 13.4*.

Naphthol AS phosphate/diazonium salt precipitates
1. Immerse the damp filter in ethanol at room temperature for the red precipitates. For blue and green precipitates, immerse the filter in ethanol that has been preheated in a water bath to 50–65°C.
2. Incubate and change the solution periodically until the color has disappeared.
3. Proceed to removal of the probe – *Protocol 13.4*

13.3.2 Stripping probe off filters

This is usually accomplished by incubating the filter at high temperature in a solution of very low ionic strength or by alkali treatment. Sometimes formamide is present to aid dissociation of the hybrids. The ease with which probe is removed will depend on the nucleic acid and on the hybridization conditions. Because of their stability, RNA:RNA and DNA:RNA hybrids require more stringent conditions for stripping than DNA:DNA hybrids. Oligonucleotide probes are usually very easy to remove. If the filter was allowed to dry between prehybridization and stripping, it may not be possible to remove the probe at all.

The methods in *Protocol 13.4* should be successful for probes labeled with either radioactive or nonradioactive probes. Start by using one of the aqueous methods, if it does not work, try raising the temperature or use a different procedure. It is easy to follow stripping of ^{32}P-labeled probes by holding a hand-held radiation monitor close to the filter. The signal should decrease to background as the stripping proceeds. The complete removal of probe can be checked by autoradiography.

There are some potential problems with reprobing filters.

- Prolonged exposure to high temperatures may lead to gradual reduction in sensitivity through loss of filter-bound nucleic acid. This happens with both loosely bound and covalently bound nucleic acid, but is more pronounced if the nucleic acid has not been covalently bound.
- Irreversible binding of the previous probe if the filter was allowed to dry.
- Incomplete removal of the previous probe even if the filter was kept damp throughout. This can give misleading results if single-

---

PROTOCOL 13.4: Stripping probe off of filters

For blots with bound DNA or RNA
1. Transfer the damp filter to 5 mM Tris-HCl, pH 8.0, 0.1 mM EDTA, 0.05% sodium pyrophosphate, 0.1 × Denhardt's solution (at least 1 ml cm^{-2}) at 65°C. Incubate 1 h with gentle agitation.
2. Discard the wash solution and repeat step 1.
3. Rinse the filter in 0.001 × SSC.

For blots with bound DNA or RNA
1. Bring a solution of 0.1% SDS, 2 mM EDTA to the boil. Pour on to the damp filter in a glass or polythene dish and allow to cool to room temperature.
2. Discard the wash solution and repeat step 1.
3. Rinse the filter in 0.001 × SSC.

For blots with bound DNA or RNA
1. Heat a solution of 65% formamide, 10 mM sodium phosphate buffer, pH 7.2, 0.1% SDS to 65°C. Immerse the filter and incubate for 1 h.
2. Discard the wash solution and repeat step 1.
3. Rinse the filter in 0.001 × SSC.

For blots with bound DNA only
1. Wash the filter twice for 10 min each time in 50 mM NaOH at room temperature.
2. To neutralize the filter remove the denaturation solution and wash the filter in about five changes of TE buffer, pH 7.5, for 5 min each time at room temperature.

Notes
1. Alkali treatment should not only denature hybrids, but should remove reporter molecules that are attached to nucleotides via alkali-sensitive bonds.
2. If hybrids persist, try repeating the procedure at successively higher temperatures.
3. After stripping, start reprobing at the prehybridization step. Alternatively store damp in a sealed polythene sleeve at 4°C, or allow to dry in air, but be very sure that stripping was successful or any remaining probe will be irreversibly immobilized.

---

stranded tails of probe remaining on the filter are complementary to sequences in the new probe. For example, if the first probe is a recombinant DNA (vector + insert) and the second is also a recombinant, but with a different insert, hybridization may occur between the two vector sequences. This will lead to false positives.

- Unsupported nitrocellulose becomes very brittle and eventually after three or four reprobings tends to fragment. This does not happen with supported nitrocellulose, nylon or charged nylon filters.

13.3.3 Reuse of the probe

Normally only a small fraction of the probe is used up, so probes can be reused until they are degraded or have decayed to too low a specific activity. To reuse the probe (now in hybridization solution), it must be denatured again by heating to a temperature above its T_m.

For aqueous solutions, this can be done by incubating in a boiling-water bath for 10 min. Formamide-containing solutions should be heated at

70°C for 30 min. The newly denatured probe can then be added to a second filter that has been prehybridized in the normal way.

References

1. **Sambrook, J., Fritsch, E.F. and Maniatis, T.** (1989) *Molecular Cloning: a Laboratory Manual.* Cold Spring Harbor Laboratory Press, Cold Spring Harbor, New York.
2. **Leary, J.J., Brigati, D.J. and Ward, D.C.** (1983) *Proc. Natl Acad. Sci. USA* **80:** 4045–4049.
3. **West, S., Schroder, J. and Kunz, W.** (1990) *Anal. Biochem.* **190:** 254–258.
4. **Conyers, S.M. and Kidwell, D.A.** (1991) *Anal. Biochem.* **192:** 207–211.
5. **Sheldon, E.L., Kellogg. D.E., Watson, R., Levensen, C.H. and Ehrlich, H.A.** (1986) *Proc. Natl Acad. Sci. USA* **83:** 9085–9089.
6. **Holtke, H.J., Ettl, I., Finken, M., West, S. and Kunz, W.** (1992) *Anal. Biochem.* **207:** 24–31.
7. **Nguyen, Q. and Heelfinger, D.M.** (1995) *Anal. Biochem.* **226:** 59–67.
8. **Thorpe, G.H.G., Kricka, L.J., Moseley, S.B. and Whitehead, T.P.** (1985) *Clin. Chem.* **31:** 1335–1341.
9. **Pollard-Knight, D., Read, C.A., Downes, M.J., Howard, L.A., Leadbetter, M.R., Phelby, S.A. *et al.** (1990) *Anal. Biochem.* **185:** 84–89.
10. **Kessler, C.** In: *Gene Probes 1. A Practical Approach* (eds Rickwood, D. and Hames B.D.). IRL Press, Oxford, pp. 93–144.
11. **Lasky, L.A., Lev, Z., Xin, J-H., Britten R.J. and Davidson, E.H.** (1980) *Proc. Natl Acad. Sci. USA* **77:** 5317–5321.
12. **Xin, J-H., Brandhorst, B.P., Britten, R.J. and Davidson, E.H.** (1982) *Dev. Biol.* **89:** 527–531.

14 Troubleshooting

However well experiments are planned, from time to time things go wrong. This chapter describes some of the more common problems that arise and suggests what may have gone wrong and how to prevent the problem occurring in the future.

The filter fell apart. This is most likely to occur with unsupported nitrocellulose filters.

- During binding of DNA to the filter, alkali was not neutralized properly. The filter would have become yellow on baking and would have been brittle. Do not use alkaline transfer buffers for Southern blotting to nitrocellulose filters. Ensure that denatured gels are properly neutralized by using excess buffer and agitating gently. Change the buffer frequently.
- The filter becomes rather brittle after repeated probings even when using the correct procedures. Prepare new filters and consider using nylon or charged nylon filters rather than nitrocellulose.

The signal is absent or lower than expected.

- Nucleic acid was not completely transferred out of the gel. After transfer, restain the gel with ethidium bromide and examine under UV light to determine if any nucleic acid was left in the gel.
- Nucleic acid did not bind to the filter. To check if there is nucleic acid on the filter, stain the filter in 0.04% methylene blue in 0.5 M sodium acetate, pH 5.2, for about 5 min. Rinse several times in 0.5 M sodium acetate, pH 5.2. (Do not do this before hybridization as it can impair the efficiency of hybridization, but it can be carried out later to diagnose problems.)

  Were the correct ionic conditions used for binding? Binding to nitrocellulose requires high ionic strength solutions. Some nylon filters require high ionic strength, some require low.

- The labeling of the probe was unsuccessful or the specific activity was too low. Check the specific activity after unincorporated nucleotides have been removed. If an end-labeled probe was used, could a uniformly labeled one have been used instead?

- The probe was degraded. This is most likely to happen with RNA probes. Treat all solutions and equipment that comes in contact with the probe to remove RNase. Use standard procedures such as in Section 10.5.1.
- RNA probes were treated with alkali. Alkali degrades RNA. RNA tends to contain intramolecular base-paired regions. These can be denatured by heat treating and quenching the tube in ice.
- The reporter molecules attached to the probe had alkali-sensitive linkages and the probe was denatured by alkali treatment. Denature by heating or use nucleotide derivatives with alkali-stable linkages to label the probe.
- The double-stranded DNA probe was not denatured. Check denaturation conditions. Work fairly quickly after denaturation so that probe has no time to renature before it is added to the filter.
- The hybridization and/ or washing conditions were too stringent so that the hybrids either did not form at all or were dissociated. Carry out pilot experiments with a set of dot blots and vary the hybridization conditions, but maintain relaxed washing conditions. Select the best hybridization conditions and then vary the washing conditions. Select best washing conditions.
- The amount of nucleic acid on the filter was insufficient to be able to detect the sequence of interest. If total RNA was used, could polyadenylated RNA be used instead?
- The hybridization time was too short.
- The filter was not exposed to film for long enough for radioactive hybrids.
- For light-based detection systems either the chemicals were degraded or the film was in contact with the filter for too short a time. There is often a lag in reaching steady-state emission of light. Was exposure too soon or too late after the substrate was added to the filter? If using a kit, consult the instructions to determine the light emission properties for the enzyme/substrate system used.
  Try exposing the filter to film for longer and if necessary repeat the chemiluminescent detection with fresh reagents.

Nonspecific background.

- The stringency of washing was too low. Try washing with a lower salt concentration or at higher temperature.
- The concentration of probe was too high. A high concentration of probe is beneficial for increasing the rate of hybridization, but too high a concentration can lead to high backgrounds. Try increasing the efficiency of labeling the probe so that less probe can be used during hybridization.

- Insufficient washing of the filters. Do not be mean with the volume of washing solutions (the chemicals required are inexpensive) and change the solutions often. Be sure that the washing solutions have been preheated to the correct temperature before they are added to the filter.
- When chromogenic substrates are used for detection, high backgrounds can occur if the wrong type of filter is used. For nitrocellulose filters, a special blocking reagent has been formulated (Tropix Inc.). Check the filter manufacturer's recommendations.
- The filter may have been allowed to dry before excess probe was washed off. Filters do not have to dry completely for probe to bind irreversibly. Even damp filters can have high backgrounds. Try keeping filters covered in solution at all times between prehybridization and hybrid detection and particularly if there is a delay between any steps.
  Try stripping and reprobing the filter (Section 13.3.2), but it may no longer be possible to remove the probe.
- The probe was 'dirty'. It may have been contaminated by traces of agarose. Either repurify the nucleic acid from which the probe was made or purify the labeled probe by one of the methods in Section 11.2. (Do not pass the probe through a nitrocellulose filter as it will stick and be lost. It can be purified by filtration through a cellulose acetate filter).
- The filter was not prehybridized properly. Or there may have been too low a volume of prehybridization solution that allowed probe to bind irreversibly. Strip and re-probe. This may be unsuccessful and a new filter may have to be prepared.
- For filters in hybridization bottles, the washing solutions may not have had access to the side of the filter next to the glass. After hybridization, remove the filters into a vessel with plenty of washing solution and change the solution frequently.
- Inadequate blocking of filters before visualizing nonradioactive hybrids. Use different vessels/sleeves for hybridizing and visualizing hybrids.

The autoradiograph of the filter is black in parts.

- Part of the filter was allowed to dry: see above.
- The filter was handled with bare hands. Grease marks from the fingers trap probe. Wear gloves in future.
- Part of the filter in a hybridization bottle doubled over, thus trapping the probe.

The autoradiograph has black dots in random locations.

- Air bubbles were not completely removed from the filter during hybridization. This causes the probe under the bubble to dry out.

Agitate the filter so that the solution flows evenly across it.
- Unincorporated radioactive nucleotides were not removed from the probe.

A 'negative' effect is obtained, i.e. the background is black with clear dots or bands.

- Too high a concentration of ^{32}P was used. Repeat with a lower concentration.

15 Specific applications of techniques

15.1 Mapping DNA

Filter hybridization is extensively used in mapping DNA. If a fragment of DNA is sufficiently long, it is likely to have recognition sites for one or more restriction enzymes and it is possible to work out their position. Mapping of cloned DNA involves digesting the DNA with restriction enzymes both singly and in combination. The digest is run on an agarose gel and from the size of bands, the position of enzyme sites can be deduced. If the DNA is blotted on to a membrane and hybridized with a probe, the position of the complementary sequence can be determined (*Figure 15.1*). Obtaining a map and locating the site of particular sequences is often necessary for characterizing a newly isolated piece of DNA.

15.1.1 Mapping repetitive sequences

In order to use a newly isolated clone as a probe for detecting additional sequences from a recombinant library, it is necessary to identify and remove any repetitive sequences. Otherwise these sequences would hybridize to every clone that contained repetitive sequences leading to a large number of false positives. There are two ways of detecting the presence of repetitive sequences in a clone. Both depend on the fact that hybridization of repetitive sequences will completely swamp the hybridization signal from single-copy sequences.

- A Southern blot of cloned DNA that has been digested in various ways is probed with uniformly labeled total genomic DNA. If repeated sequences are present in the clone, the fragments containing the sequences will give a hybridization signal. If no repeated sequences are present, there will be no signal.
- A Southern blot of digested genomic DNA is hybridized with labeled, cloned DNA. If the clone contains repetitive sequences, the

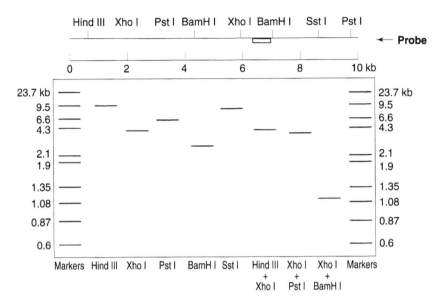

Figure 15.1. Determining the position of a sequence of interest in a cloned fragment of DNA. The DNA is restricted in single and double digests and the products are separated by gel eletrophoresis. A Southern blot of the DNA is probed with the sequence of interest. The bands which hybridize are correlated with the restriction map to localize the sequence to the smallest fragment. The procedure can be carried out in conjunction with restriction mapping of the DNA on an ethidium bromide gel or once the map is known.

hybridization signal will appear as a smear because there will be many genomic DNA fragments carrying repeated sequences and these will have a large range of sizes. If the clone lacks repetitive sequences, distinct bands will be observed at the positions of complementary sequences.

Once the position of the repeated sequences is known it is usually easy to remove them by restriction digestion and to reclone the sequence of interest. However, this can be tedious and alternative strategies for avoiding the complications of repetitive sequences are discussed in Section 15.6.

15.2 Detection of related sequences

15.2.1 Discrimination between related sequences

A sizeable proportion of the eukaryotic genome is composed of families of similar, but not identical, sequences, for example globin, immunoglobulin and *ras* genes. It is often the aim of filter hybridization to distinguish between closely and distantly related members of such families, for example screening recombinant libraries and determining the size and copy number of related sequences on Southern blots (*Figure 7.1*).

Experimentally, related family members are treated as sequences that differ in their degree of homology to the probe. So, hybridization and washing procedures are modified to tolerate mismatching. It will not be known in advance how near or distantly related the family members are, so optimal hybridization and washing conditions have to be determined by trial and error. The problem becomes one of finding hybridization conditions that will allow mismatched duplexes to form and washing conditions that are mild enough to retain the hybrids, but stringent enough to remove nonspecific background.

Pilot experiments are carried out with replicate strips from a Southern blot. Each is hybridized with the same probe at different stringency and washed under identical relaxed conditions. The hybrids formed are visualized. The filters are then washed at increasing stringency and the hybridization signal is monitored each time. The experiment can be repeated at higher and higher stringency of hybridization until the conditions for detecting perfectly matched hybrids are reached.

To detect well-matched hybrids, use stringent hybridization followed by a stringent wash. This is better than permissive hybridization followed by a stringent wash [1]. For detecting more distantly related sequences use the most appropriate relaxed criteria as determined in the pilot experiments above. Note that a single compromise criterion is not effective at discriminating between closely and distantly related sequences. Too high a hybridization temperature will mean that poorly matched hybrids are not stable. Too low a hybridization temperature will mean that the rate of formation of well-matched hybrids will be very slow. This can not necessarily be compensated for by prolonging hybridization.

It might appear that the longer the hybridization, the better the discrimination, but this is not the case. The extent of hybridization is an important factor in discriminating between sequences. The incubation time for greatest discrimination depends on whether the filter-bound or probe sequences are in excess [1]. In practice it is best to use short incubation times regardless of whether the probe or filter-bound sequences are in excess and if this does not generate enough signal to be detected, excess filter-bound sequence should be used and the hybridization time extended.

15.2.2 *Zoo/garden blots*

The sequence of coding DNA is strongly conserved in evolution whereas noncoding sequences are not under the same selection pressure and diverge fairly rapidly between species. Coding sequences from one species can be used to identify putative coding species from another by hybridizing under conditions of reduced stringency.

Zoo blots are Southern blots containing restriction enzyme digests of DNA from a variety of different animal species. Similarly, garden blots contain DNA from different plant species. These blots may be prepared in-house, but are also available commercially. A limitation of in-house blots is likely to be obtaining DNA from the desired species. Disadvantages of commercial blots are that they are expensive and the range of restriction enzymes used may be limited.

Since samples are precious and may not be available again, the DNA is bound covalently to a nylon or charged nylon filter. This enables the filter to be stripped and rescreened repeatedly without loss of DNA. A variety of stringencies will have to be tested as different species will probably have different degrees of homology with the probe.

15.3 Restriction fragment length polymorphisms

The vertebrate haploid genome contains about 3×10^9 base pairs, so it is perhaps not surprising that there is much variability in DNA sequence between unrelated individuals of the same species. If differences at a certain location occur in a high proportion of the population, they are known as polymorphisms rather than mutations. Many polymorphisms are located in introns or between genes and have no phenotypic effect.

Some polymorphisms can be detected by Southern blotting. Electrophoresis separates restriction fragments on the basis of size, if there is a moderately sized insertion or deletion of DNA, the size of restriction fragment will be altered. Similarly, if a point mutation creates or abolishes an enzyme recognition site, the size of restriction fragment will be altered. These changes are known as restriction fragment length polymorphisms (RFLPs).

Polymorphisms are inherited in a simple Mendelian fashion and can be exploited in gene tracking. An example is shown in *Figure 15.2* which sought to determine if a girl was a carrrier of the condition muscular dystrophy.

Muscular dystrophy is an X-linked disease caused by mutations (usually deletions) in the dystrophin gene. A polymorphism near the 5′ end of the gene affects a *Taq*I recognition sequence and is detected as a change in size of a *Taq*I band on an audioradiogram. The polymorphism has nothing to do with causing the disease but is situated sufficiently close to the gene that it is inherited together with the gene. Determining the inheritance pattern of the dystrophin gene by RFLP analysis depends on being able to correlate one of the *Taq*I bands with the occurrence of disease so that inheritance of the RFLP implies inheritance of the mutated gene.

In *Figure 15.2*, the affected boy (lane 3) has inherited the upper RFLP from his mother whereas the unaffected brothers have inherited the lower RFLP. (The boys do not inherit a dystrophin gene from their father because it is on the X chromosome.) The father has given his daughter the lower RFLP, so by difference the mother has contributed the upper RFLP. This is the same RFLP that was inherited by the affected brother, so the girl is a carrier of this X-linked disease.

RFLP analysis can also be carried out when the gene responsible for a disease is unknown, but its approximate position on the chromosome

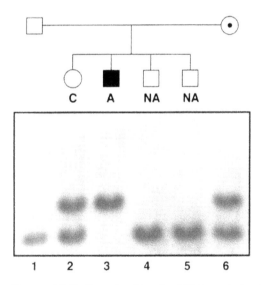

Figure 15.2. Gene tracking by RFLP analysis. A single base change linked to the 5′ end of the dystrophin gene affects a *Taq*I recognition site. The RFLP can be detected by the change in position of a *Taq*I band on an autoradiogram. To determine the inheritance pattern of the X-linked disease, a Southern blot of *Taq*I restricted DNA from father, mother and four children is probed with XJ1.1 which detects the RFLP. The boys have inherited their single band from their mother while the girls have inherited one band from each parent. By comparing the patterns of parents and children it can be deduced that the affected boy has inherited the mutation from his mother and that the daughter is a carrier of muscular dystrophy.

has been located. It is necessary to find a polymorphism close enough to the gene so that the gene and the polymorphic site are inherited together and recombination at meiosis rarely separates them.

A pedigree analysis of family members similar to that above is carried out to determine if the presence/absence of the RFLP is correlated with the occurrence of disease. If correlation exists, then prenatal DNA from chorionic villi can be examined for the presence of the polymorphism and the result can be used as a strong indicator of carrier or disease state. The analysis will fail if both parents in an affected family are heterozygous for the polymorphism because it will not be possible to tell from which parent the mutant gene was inherited. This illustrates the main disadvantage of RFLP analysis – there are only two alleles (possessing or lacking the restriction site polymorphism) and three possible states (homozygous for the presence/absence of the polymorphism and heterozygous).

15.4 DNA fingerprinting by hybridization

Another major source of variation in DNA sequence between individuals is the highly polymorphic minisatellite sequences also known as variable **n**umber of **t**andem **r**epeats (VNTRs). These are short stretches of DNA typically 9–30 nucleotides in length, that are repeated many times in tandem. The polymorphic repeat size ranges from about 0.1 to 20 kb. Many of these repeats have evolved from a common ancestor and share the core sequence of -GGGCAGGAXG- where X represents any nucleotide. In humans there are over 1000 hypervariable minisatellites, most of which are found near telomeres.

That VNTRs share sequence similarity can be shown by using any VNTR to probe a Southern blot at low stringency of hybridization. These conditions permit imperfectly matched sequences to form hybrids.

Jeffreys [2,3] used two VNTR probes, 33.15 and 33.6, to probe DNA that had been cut by the enzyme *Hinf*I. Many bands gave a hybridization signal. The number and size of the bands obtained was specific for each individual. These polymorphic fragments are stably inherited and segregate in a Mendelian fashion. Because they are stable and specific to an individual they are widely used as the basis of identity tests. In conjunction with statistical analysis they are also used for establishing the relatedness between individuals. They have also became useful tools in disputes over proof of kinship (*Figure 15.3*) and in forensic analysis.

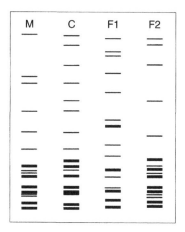

Figure 15.3. A hypothetical paternity case confirmed by fingerprinting. A Southern blot of restricted DNA from mother (M), child (C) and two possible fathers (F1 and F2) is probed with a multilocus probe. The fingerprint band pattern of the child must be a composite of that of its parents, so by comparing the patterns from mother and child, the band pattern of the father can be deduced. Only F2 can be the father.

Such probes are known as **multilocus probes** (MLPs) because they detect many different loci simultaneously. The pattern of bands became popularly known as bar codes and the process as DNA fingerprinting. Useful though they are, multilocus probes do have disadvantages. For example, a minimum of about 0.1–1 µg DNA is required for analysis. The accuracy of scoring bands is very important particularly with faint bands because an error in scoring may affect the outcome of a legal case. Another disadvantage is that it is not known which bands in the fingerprint are alleles.

Single locus probes (SLPs) are now much used in DNA profiling. (DNA profiling describes the screening of DNA for variation whereas DNA fingerprinting is reserved for the technique devised by Jeffreys.) Here the variability of DNA is measured at just one locus at a time so that only two bands are detected, one from each allele (*Figure 15.4*). High stringency of hybridization is used to confine the hybridization to the desired target. In practice five different targets are separately probed using unlinked probes. The loci chosen are highly polymorphic (a large number of possible variants is possible) and a large proportion of the population are heterozygous at these sites. Statistical analysis is used to show the frequency of occurrence of the pattern of alleles in various populations such as Caucasian or Afro-Caribbean. Each separate probe will not allow a unique identification of an individual, but if several SLPs are used that have been rigorously tested, then

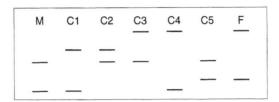

Figure 15.4. A hypothetical family relationship confirmed using a single locus probe. Each of five children (C1–5) inherits one allele from its mother (M). The younger three children (C3–5) all inherit an allele from their father (F). The elder two children (C1–2) inherit an allele from their deceased father – the mother's first husband. In reality five different single locus probes would be used to confirm the family relationship.

the cumulative frequency of occurrence of all the variables can give a unique identification.

Much DNA profiling is now carried out by PCR which is faster, can start with much smaller amounts of material, can assay all five loci in a single reaction and can score polymorphisms automatically [4,5].

15.5 Detection of mutations

Detection of mutations is extensively used in basic research and is particularly useful in medical diagnosis.

15.5.1 Loss of enzyme restriction sites

A change in the sequence of an enzyme recognition site is called a restriction site polymorphism (RSP). Such a mutation abolishes enzyme activity at that site giving rise to an RFLP. This presents an easy means of detecting the mutation on Southern blots.

Some pathogenic mutations destroy enzyme recognition sites. Sickle cell anemia is a genetic disease in which a single base change of an A to a T in codon 6 of the β-globin gene destroys an *Mst*II site. The flanking *Mst*II sites are located 1.2 kb upstream and 0.2 kb downstream. Normal DNA will generate fragments of 1.2 and 0.2 kb when digested by *Mst*II, whereas DNA containing the sickle cell trait will generate a fragment of 1.4 kb. These can be distinguished by Southern blotting and hybridizing DNA with a β-globin specific probe (*Figure 15.5*).

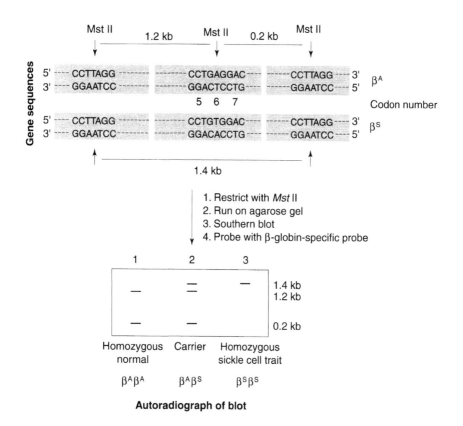

Figure 15.5. Detection of a mutation by loss of an enzyme recognition site. The mutation that causes sickle cell anaemia (the carrier or sickle cell trait) is detected by the altered position of DNA bands on an autoradiogram.

Some pathogenic changes that occur in the onset of cancer can also be detected easily because they alter enzyme restriction sites. For example an early change in the onset of certain cancers, creates an *Hph*I recognition site in codon 13 of K-*ras* gene. Similarly, pathogenic mutations in codon 12 of H-*ras*1 abolish an *Msp*I recognition site. These mutations can be detected by comparing restriction patterns of DNA from paired normal and tumor tissue from the same individual.

With the large variety of endonucleases commercially available, it is surprising how often an enzyme can be found which recognizes either the normal or mutant sequence. However, some of the rather obscure enzymes are expensive and not very active, so alternative methods of detecting mutations may be preferable.

15.5.2 Loss of genes

Deletion of certain genes, tumor suppresser genes, contributes to the development of some cancers. Southern blotting can be used to screen for deletion of known tumor suppresser genes in paired blood and tumor samples from a patient. Deletions can involve one or both alleles. If both alleles are deleted, the band will be missing whereas if only one allele is deleted, the intensity of the hybridization will be half that in the normal sample. This should be obvious to the naked eye provided that the autoradiograph has not been overexposed, but can be confirmed by scanning the autoradiograph.

15.5.3 Expansion of repeated sequences

Genomes of many eukaryotes contain many simple sequences that are repeated in tandem and dispersed throughout the genome. The number of repeats in a unit is inherited in a Mendelian fashion and is generally stable. In humans it has been shown that certain trinucleotide repeats are stable if the length falls below a certain threshold, but above it the repeat length is unstable and becomes increasingly larger with increasing number of mitosis or meiosis. Expansions of triplet repeats underlie diseases such as myotonic dystrophy (CTG), Huntington's disease (CAG) and the most common form of Fragile X syndrome (CGG).

Detection of small-scale expansions can be made either by PCR (polymerase chain reaction) and gel electrophoresis or by Southern blotting using a probe flanking the triplet repeat. Some of the expansions are so massive that they are too big to be amplified by PCR and detection has to be by Southern blotting.

The cause of myotonic dystrophy is thought to be an expansion of the triplet CTG in the 3′ untranslated region of a protein kinase gene (DMPK). It has been found that the extent of expansion generally correlates with the age of onset and severity of the disease. The normal range of repeats lies between 5 and 35: mild clinical symptoms are found with 50–150 repeats: classical myotonic dystrophy with 100–1000 repeats and congenital disease with over 1000 repeats. So knowledge of the degree of expansion can be useful in managing the condition.

The autoradiograph in *Figure 15.6* shows a Southern blot analysis of the trinucleotide repeat in myotonic dystrophy. The control in lane 1 has five trinucleotide repeats on both X chromosomes. The mother (lanes 2 and 3) who has a mild form of the disease has the normal five repeats on one chromosome, but an expansion (90 repeats) on the other. The son (lane 4) has one normal allele and about 350 repeats on the other. The daughter (lane 5) has one normal allele and about 500 repeats on the other. Both

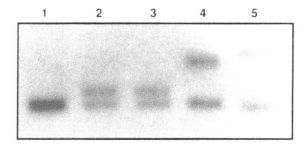

Figure 15.6. Expansion of the trinucleotide sequence underlying myotonic dystrophy detected by Southern blotting. A Southern blot of *Bgl* I digested lymphocytic DNA probed with pFB1.4 which flanks the repeat. Lane 1, normal control. Lanes 2 and 3 the mother, Lane 4, the son and Lane 5 the daughter.

children have classical myotonic dystrophy. The 'fuzziness' of the upper bands in the children probably reflects a range of expansions in the lymphocyte population from which the DNA was derived.

15.5.4 Use of allele specific oligonucleotide probes

The presence of a single mismatched base decreases the T_m of an oligonucleotide-containing hybrid by 5–10°C. The extent of the decrease depends on the length of the oligonucleotide and the position and identity of the mismatched base. Under carefully optimized incubation conditions, an oligonucleotide will hybridize only to its fully complementary sequence and not to one that differs by a single base. This is the basis of a method called allele-specific oligonucleotide hybridization [6,7] that is extensively used to screen sequences for known mutations (*Figure 15.7*).

Two oligonucleotide probes are synthesized, one being complementary to the normal (wild-type) sequence and the other to the mutant sequence. So each oligonucleotide is specific for one allele. The oligonucleotides have the same sequence except for the mutant base which is as near the center as possible.

Two identical blots are prepared which contain the samples being screened for the mutation and three controls:

- DNA that is homozygous for the normal sequence;
- DNA that is homozygous for the mutant sequence;
- DNA that is heterozygous, i.e. one allele is normal and the other is mutant.

One blot is hybridized with the oligonucleotide probe having the normal sequence whereas the other is hybridized with probe having the mutant sequence. The hybridization and washing temperatures are chosen to be

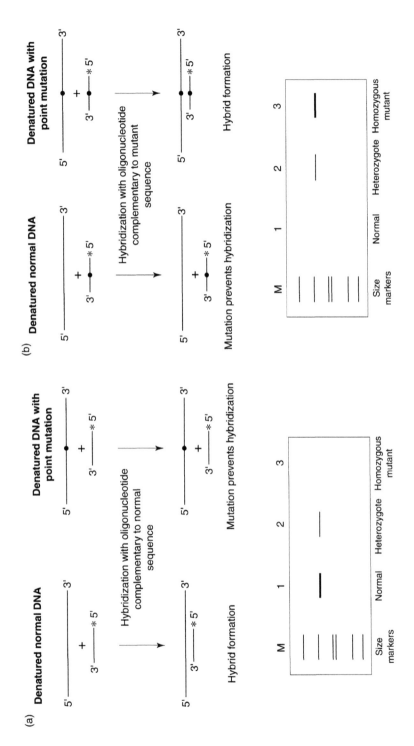

Figure 15.7. Detection of a mutation using allele specific probes. Duplicate Southern blots are probed with oligonucleotides probes complementary to (a) the normal and (b) the mutant sequence. Probes are labeled at the 5′ end with ^{32}P. The probes have the mutant base in the middle so that mismatched hybrids are thermally unstable and give no radioactive hybridization signal. The intensity of hybridization signal on the autoradiograph distinguishes the presence of two normal sequences from one normal and one mutant sequence.

no more than 5°C below the T_m for perfectly matched hybrids. These conditions are stringent and are designed to prevent mismatched hybrids from forming. The blot hybridized with the normal probe will only give a hybridization signal if the normal sequence is present and conversely, the blot hybridized to the mutant probe will only give a signal if the mutant signal is present. It is necessary to be sure that absence of a hybridization signal is caused by absence of the complementary sequence and not to some trivial cause such as inappropriate incubation conditions. This is why positive controls are included.

The design of the pair of oligonucleotide probes and the hybridization conditions used are critical. There must be the greatest possible difference in thermal stability between the perfectly matched and mismatched hybrids. The variables that can be manipulated in designing the oligonucleotide are the length of oligonucleotide and the position and identity of the mismatched base.

- The oligonucleotides must be long enough to ensure that they hybridize only to the desired target sequence, but they need to be relatively short to maximize the difference in stabilities of the hybrids. The longer the oligonucleotides are, the lower will be the depression of T_m by mismatching. In practice, probes of 17–20 nt are usually used for identifying mutations.
- The mismatched base is most disruptive if it is in a central position, but in any case it should be at least five bases from either end.
- The choice of sense or antisense strand as target for the probes may be important. As discussed in Section 8.2.2, the identity of the mismatched base has a considerable effect on the stability of a hybrid. The target strand should, therefore, be that for which the mismatched base renders the heteroduplex most unstable.
  The target bound to the filter can be RNA rather than DNA. In this case, the probe must be an antisense oligonucleotide. If this does not discriminate sufficiently between normal and mutant sequences, a cDNA can be prepared and screened with a sense oligonucleotide.

For hybridization, conditions must be found that allow formation of perfectly matched hybrids, but prevent the formation of mismatched ones. The T_m values for perfectly matched normal and mutant hybrids can be estimated from Equations 8.1 or 8.2. Pilot experiments are carried out using hybridization temperatures that range from about 2 to 10°C below the calculated T_m. Washing should be stringent: 1–2 min at 5°C below the T_m followed by a wash at room temperature to remove probes that have dissociated from the target DNA.

Probing with allele-specific probes is widely used in prenatal diagnosis of disease. The amplification afforded by PCR and discrimination afforded by allele-specific oligonucleotide hybridization can be combined to screen mutations quickly and accurately [8]. DNA spanning the site of the mutation is amplified by PCR from paired normal and test DNAs. The amplified DNA is dot blotted on to filters and hybridized with allele-specific probes. The advantage of this system over hybridization alone is that the target sequence is available in much higher amount. The advantage over allele-specific PCR alone is that it is easier to optimize the hybridization reaction to give maximal discrimination between the mutant and normal sequences.

15.5.5 Reverse dot blots

In conventional dot blots, a variety of different samples can be probed at once, but with one probe at a time. If several sequences are to be detected, then multiple hybridizations must be carried out. Reverse dot blots were invented to enable the simultaneous detection of a number of different sequences in a single hybridization [9,10]. The method is simple. An array of different probes is bound to a filter and is hybridized to a single labeled test sample.

One of the first uses of the technique was the identification of a pathogenic micro-organism present in a clinical sample [9]. A dot blot was prepared containing DNA from those microorganisms that are most commonly found in urinary tract infections. DNA from the clinical sample was labeled and hybridized to the panel of DNAs on the filter. The hybridization results identified a micro-organism and agreed completely with results obtained by conventional clinical microbial analyses.

15.5.6 Ligase-mediated gene detection

The oligonucleotide ligation assay provides an alternative to allele-specific oligonucleotide hybridization for detection of mutations in DNA. The principle of the method is that two adjacent oligonucleotides annealing to the same template can be ligated together if the ends facing each other are perfectly base-paired to the template, but not if there is a gap between them or if the oligonucleotides are incorrectly base paired at the junction. The presence of a mutation in the template DNA at the junction prevents ligation and thus can be detected.

In practice, biotin is attached to the 5′ end of one oligonucleotide and a suitable reporter, such as ^{32}P, to the 3′ end of the other (*Figure 15.8*). The oligonucleotides are annealed to denatured DNA in solution and incubated with DNA ligase. After ligation, the biotin-carrying oligonu-

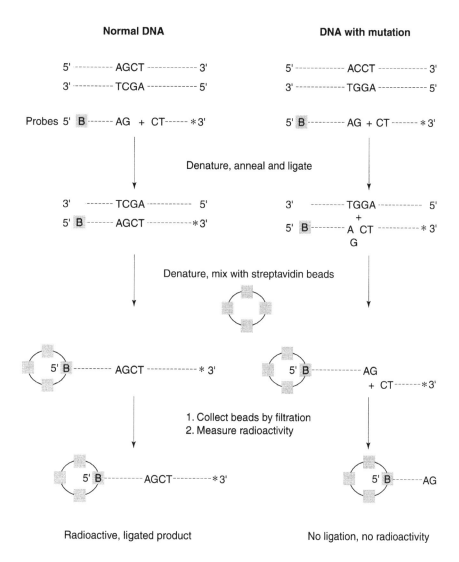

Figure 15.8. Detection of mutations by the oligonucleotide ligation assay Target DNA is denatured by heating and mixed with two oligonucleotides that lie adjacent to each other in head to tail fashion. In the presence of DNA ligase the oligonucleotides will be ligated only if they match the target sequence perfectly at the junction. The oligonucleotides are labeled with different labels so that only ligated molecules will carry both labels.

cleotides are isolated on a streptavidin-coated surface. If the biotin-containing molecules are radioactive, ligation has taken place. Ligation is measured by autoradiography. The assay is fast and also has the potential for automated detection if the second reporter molecule is a fluorophore. The products of ligation can be loaded on an automated gel sequencer and analyzed much faster than by using autoradiography. [11].

The principle of the ligase chain reaction (also known as ligase amplification reaction (LAR) is similar to the oligonucleotide ligation assay wherein oligonucleotides correctly positioned on the target DNA are ligated. But here the yield of ligated product is increased by using repeated cycles of ligation and denaturation with a thermostable DNA ligase [12,13].

15.6 Repeated sequences in probes

15.6.1 Suppressing hybridization of repetitive sequences

Most eukaryotes contain repeated sequences. Some are clustered and others are dispersed throughout the genome. The proportion of the genome occupied by these sequences varies widely. In humans, the *Alu*I repeat occurs on average once every 4 kb and the LINE-1 or (L1 or *Kpn*I) repeat occurs on average once per 50 kb.

The number and spacing of these sequences is such that there is a high probability that cosmid, P1 and YAC bacteriophage λ recombinants will contain a repeated sequence. If inserts from these recombinants are used as probes on genomic Southern blots, they will produce a smear which will obscure the signal from single-copy elements. This is because repetitive elements having low complexity and high concentration on the filter, hybridize faster than single copy sequences (i.e. repetitive elements hybridize at low C_0t values whereas single-copy sequences hybridize at high C_0t values).

There are three ways to overcome this problem.

- The single-copy sequence can be subcloned out of the recombinant probe. This will involve characterizing the probe and is likely to be tedious.
- A quicker and easier way is to suppress the ability of repetitive sequences to hybridize to filter-bound targets [14,15]. This can be done by preincubating the labeled probe in solution with excess genomic DNA for a time that has been calculated to allow

repetitive, but not single copy hybrids to form. This effectively sequesters the repetitive elements in the probe in double-stranded form so that they will be unavailable to hybridize to filter-bound sequences. Since the single-copy sequences in the probe will not have had time to hybridize, they will remain single-stranded and will be able to hybridize to the target. There is no need to remove the hybrids from the solution containing the prehybridized probe which is then added to the filter and hybridization is carried out as usual.

A large excess (1 mg) of unlabeled competitor or 'driver' DNA is sheared or sonicated to about 500 nt and mixed with 10–100 ng of labeled probe in a small volume of liquid (100 μl of 5 × SSC). The mixture is heated to denature the nucleic acid and is then incubated at 68°C to a C_0t value of between 10 and 100 (mg ml^{-1} × min) [14]. This is below the $C_0t_{1/2}$ of single copy DNA. The mixture is then added to a filter containing target sequences that has been prehybridized in the normal way and hybridization is carried out as usual.

C_0t can be calculated using the relationship:

C_0t = 'driver' DNA concentration (mg ml^{-1}) × time (min)

where 'driver' is the competitor DNA that is present in excess.

A detailed discussion of prehybridization strategy can be found in ref. [14].

- The easiest way to overcome the problem of repetitive sequences on the probe is to prehybridize the probe in solution with unlabeled DNA that has been enriched in repetitive sequences (so-called Cot-1 DNA). When excess Cot-1 DNA is denatured and prehybridized to the probe, the repetitive elements in the probe are sequestered in a double-stranded form so that they are prevented from hybridizing to blots The background on filters obtained with Cot-1 DNA is lower than that obtained with total genomic DNA.

  Cot-1 DNA is prepared by fragmenting and denaturing DNA then reassociating in solution to a C_0t value of 1 [16]. The reassociated component which is composed of repetitive DNA can be isolated as described in Section 3.5. Cot-1 DNA is also available commercially, for example from Life Sciences who sell Cot-1 DNA from several species.

15.6.2 *Using repetitive sequences as probes*

The Alu sequence in humans is similar in sequence to the B1 family of repeated sequences in the mouse, but is sufficiently different that conditions can be found where they do not hybridize to each other. This is very useful as Alu probes can be used to screen libraries from human–mouse hybrids to identify clones containing human inserts.

15.7 Semi-quantitative analyses

In order to quantitate the amount of DNA or RNA in a sample by hybridization, it is necessary for the reaction to go to completion. Otherwise the method will underestimate the number of species present and their prevalence. Solution hybridization is ideally suited for quantitative studies. Sequences present in high concentration enter duplexes quickly, but for rare sequences it may be necessary to prolong incubation for up to 3 days to ensure that all potential hybridizing sequences have become double-stranded.

Filter hybridization is about 10 times slower than solution hybridization [17] which means that it is very difficult to reach the high $C_o t$ values necessary for rare sequences to hybridize. In addition, attachment to a filter prevents some target sequences being available for nucleation. For these reasons, filter hybridization is not used for quantitative measurement of nucleic acid.

However, filter hybridization can be used semiquantitatively. Dot blots can be used to compare the relative abundance of different nucleic acid species and are particularly useful when many samples are to be analyzed at once.

An example of a very simple semiquantitative analysis of several genes in genomic DNA is shown in *Figure 15.9*. Replicate filters of serial

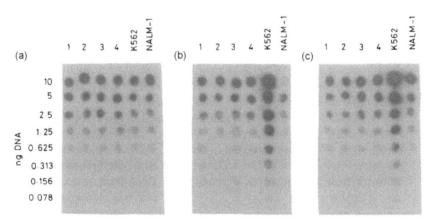

Figure 15.9. Semi-quanatitative analysis of DNA dot blots. Replicate nitrocellulose filters containing the indicated amounts of genomic DNA from a dilution series were probed with ^{32}P-labeled probe. (a) c-*sis* oncogene. (b) c-*abl* oncogene. (c) IgV$_\lambda$ DNA. Lanes 1–4 contained DNA from peripheral blood of chronic myeloid leukaemia (CML) patients, 5 and 6 contained DNA from CML cell lines K562 and NALM-1 respectively.

dilutions of genomic DNA were prepared. The DNA was derived from patients with chronic myeloid leukemia (CML) and two CML cell line DNAs. Each filter was hybridized with a different ^{32}P-labeled probe. The probe was present in excess over filter-bound sequences so that the extent of hybridization was a measure of the concentration of hybridizing species on the filter. It can be seen that the cell line K562 contains about four times more copies of the oncogene c-*abl* and immunoglobulin V_λ genes than the other cell samples. The amplification is not shared by the c-*sis* oncogene.

Similar analysis can be carried out with RNA dots. By using a calibration curve, the extent of hybridization of the probe to filter-bound sequences can be used to determine the actual number of transcripts [18]. The method involves hybridization in solution to establish the copy number of transcripts per ng of mRNA for several RNAs that are present at widely different abundances. This allows a calibration curve to be prepared relating number of transcripts per ng mRNA to radioactivity per dot. From the graph, the number of transcripts of any other recombinant on the matrix can be obtained. For a more detailed account, see [19].

15.8 Methylation

Covalent modification of DNA by attachment of methyl groups is an important mechanism for controlling gene expression. Between 3% and 5% of cytosine residues in vertebrates are methylated with the vast majority being present in the CpG dinucleotide.

Southern blots can be used to study methylation patterns in DNA. The method depends on the ability of certain enzymes to cleave at their recognition site whether or not it contains a methylated base whereas isoschizomers will only cut if the sequence is unmethylated. The restriction enzymes *Msp*I and *Hpa*II are widely used for determining methylation status of DNA. The recognition sequence for both is CCGG, but *Hpa*II will not cut if the sequence is methylated.

Experimentally, duplicate samples of genomic DNA are treated with either the methylation-insensitive or the methylation-sensitive enzyme. Southern blots of the digests are probed for the target sequence (*Figure 15.10*). If the pattern of digestion is identical, then the DNA is unmethylated. However, if the pattern is different, the sequence has been methylated.

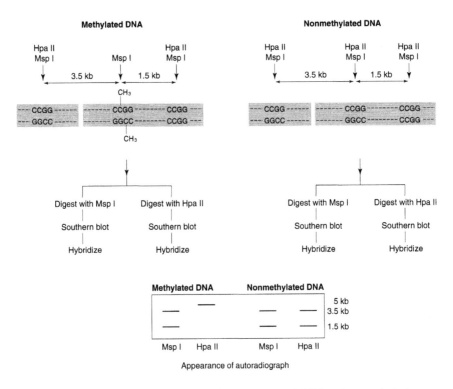

Figure 15.10. Determination of methylation status of DNA using methylation-sensitive restriction enzymes. Duplicate samples of DNA are incubated with restriction enzymes *Hpa*II and *Msp* I. Restriction by *Hpa*II is sensitive to the presence of methyl groups. A Southern blot of the digests is probed with the sequence of interest. The hybridization pattern will be the same if the enzyme recognition site is unmethylated and different if the DNA is methylated. Positive and negative controls are required to check that the enzymes are active.

This approach is useful, but is limited by the fact that there are insufficient enzyme pairs to screen all potential methylated sites.

15.9 Analyses of transcripts

Characterizing transcripts is a very important aspect of gene analysis. The primary transcript of a gene is initiated at the first nucleotide and terminates downstream of the eventual polyadenylation site. Processing generally involves adding a 'cap' to the 5′ end, excision of introns and excision of 3′ sequences followed by polyadenylation.

Characterizing transcripts usually involves determining the 5' and 3' termini, the number of introns, alternative splicing sites and possibly the existence of related transcripts. Hybridization-based techniques are invaluable in these analyses and are described below [20–22].

15.9.1 Preliminary characterization of transcripts

If both cDNA and genomic DNA clones are available, a preliminary comparison of their restriction digest pattern will provide information on the number and location of introns. This can be substantiated by probing a Southern blot of restricted genomic DNA fragments with labeled cDNA. However, in many instances the cDNA clones are incomplete so mapping information is usually incomplete.

Northern blots. Genomic or cDNA probes can be used to probe Northern blots. With processed RNA, the most intense and probably smallest band should be that of mature mRNA. With total or nuclear RNA, the presence of primary and partially processed transcripts may be detected. By varying the restriction fragment of DNA used as a probe, the position of exons and introns can be confirmed. If alternative splice sites are used in different tissues, they may be revealed by analyzing total RNA from different tissues on a Northern blot.

'Run-on' analyses. The position at which transcription terminates can be determined by 'run-on' transcription followed by Southern or dot blotting. In this procedure, isolated nuclei are incubated *in vitro* in the presence of labeled NTPs. Under these conditions there is little or no *de novo* synthesis, but RNA molecules that had been initiated *in vivo* are completed although not processed. The products of run-off are used as probes on blots that contain different restriction fragments of genomic DNA (*Figure 15.11*). The transcription end point is not unique but occurs over a stretch of about 1 kb.

Nuclease S1 protection with unlabeled DNA. Transcripts can be mapped by nuclease S1 protection (*Figure 15.12*). Unlabeled genomic DNA is hybridized in solution to RNA. Primary transcripts will hybridize along the length of the DNA whereas mRNA will hybridize to exons leaving the introns looped out. The hybrids are treated with nuclease S1 under conditions where single-stranded nucleic acids are preferentially digested. This treatment removes unhybridized DNA and RNA, overhanging 5' or 3' tails of hybrids and looped out introns, so that after digestion, only completely double-stranded duplexes remain. The products are run on native and denaturing gels and transferred to a filter. Hybridization with a labeled genomic probe reveals the products of nuclease S1 treatment.

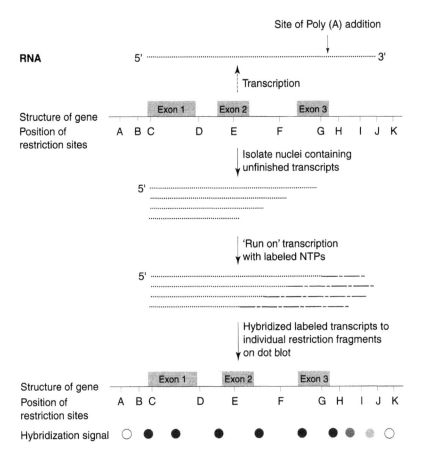

Figure 15.11. RNA mapping – Run on transcription. Mapping the site at which transcription terminates. Nuclei are isolated and transcription is completed *in vitro* using labeled nucleotides. The products are hybridized to dot or Southern blots containing isolated restriction fragments of DNA. No hybridization occurs upstream of the gene. The hybridization signal is high within and just downstream of the gene and decreases in the fragments in which transcription finishes.

On a native gel, the primary transcript gives rise to a single protected band that is double-stranded all along its length and smaller than the DNA fragment added initially. The size of the protected band on the denaturing gel will be the same as on the native gel. The processed RNA will also give rise to a single protected band on the native gel, but the hybrid will contain nicks at the positions of the introns. These will be revealed on the denaturing gel where the number and size of bands should exactly match the number and length of exons. By hybridizing the blot with different restriction fragments, the identity of the exons can be determined.

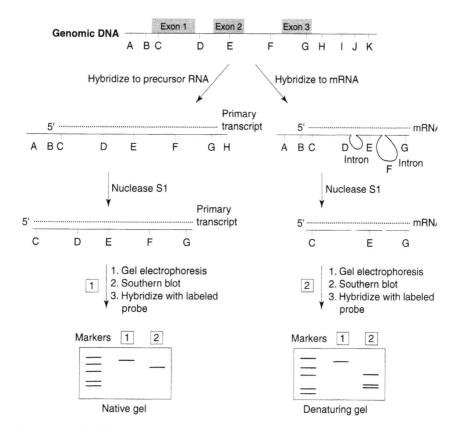

Figure 15.12. RNA mapping – Nuclease S1 protection with unlabeled nucleic acids. Determining the size of a primary transcript and the number and size of exons. Genomic DNA is hybridized in solution to precursor RNA or mRNA in the presence of a high concentration of foramide to prevent DNA:DNA hybridization. On nuclease S1 digestion, single-stranded regions of nucleic acid are removed. The products of digestion are separated on native or denaturing gels and blotted on to a filter. The position of the protected bands is located by hybridization with a labeled probe followed by autoradiography. The length of the primary transcript is the same on both native gels. The size of the mRNA is obtained from the native gel, and the number and size of exons from the denaturing gel. The identity of the exons can be established by hybridizing the blot to different restriction fragments.

This method is quick and very useful for producing preliminary mapping data. A single DNA fragment containing multiple exons can be roughly mapped in a single experiment. The mapping is not precise, however because agarose gels are usually used and the position of size markers gives only an approximate size of the protected fragment.

RNase protection assay. The RNase protection assay is a highly sensitive procedure for detecting and mapping specific RNA sequences using RNA probes (*Figure 15.13*).

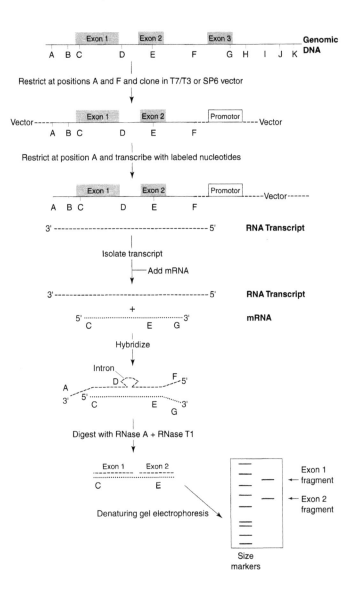

Figure 15.13. RNA mapping – RNase protection. Determining the number and size of exons. Highly labeled RNA probe is prepared by run off transcription of cloned DNA. Labeled RNA is hybridized to mRNA in solution and the RNA:RNA hybrid is treated with RNase to remove single-stranded sequences. The products are resolved on a denaturing polyacrylamide gel and autoradiographed.

The method requires that DNA spanning the region of interest be cloned into a specialized vector containing bacteriophage promoters (as in *Figure 11.3*). A uniformly labeled antisense RNA probe is transcribed from the cloned DNA using radioactive NTPs and hybridized to total or poly(A)⁺ RNA. The RNA:RNA hybrid is incubated with a mixture of single-strand specific ribonucleases [RNase A (C and U specific) and T1 (G specific)] to generate a double-stranded fully protected fragment. The protected fragment is denatured and analyzed on a denaturing gel. Radioactive bands with the size of the exons are detected by autoradiography.

The RNase protection assay will give the sizes of protected species, but not their co-ordinates. However, it is very sensitive and is particularly useful for detecting rare transcripts. The high sensitivity arises because large amounts of probe are synthesized which is useful in driving the hybridization towards completion, RNA:RNA hybrids are more stable than DNA:RNA hybrids and being single-stranded, the probe does not reassociate.

The assay is also useful for quantitating specific RNA species and detecting alternative exon use. There are two main disadvantages with the RNase protection assay. First, in order to generate the labeled probe, the DNA has to be cloned into an appropriate vector. This can be time consuming, but is useful if multiple analyses have to be made from the same DNA fragment. Second, digestion with RNase is base-specific so not all regions that are single-stranded are removed.

15.9.2 *Refined mapping of transcripts*

Once a rough estimate of the position of termini and exon–intron junctions has been made, more precise mapping can be carried out. The methods used require hybridization in solution with end-labeled probes and enzyme digestion of hybrids to reveal the termini. They are technically demanding and depend on careful optimization of all the steps.

Nuclease S1 mapping with end-labeled probes. This procedure is a variation of the nuclease S1 method discussed above.

RNA is hybridized to a molar excess of single-stranded, end-labeled DNA that is complementary to the RNA over only part of its length (*Figure 15.14*). After hybridization, the mixture is treated with nuclease S1 under conditions which preferentially digest single-stranded nucleic acid and leave double-stranded DNA intact. That part of the probe having no complement is digested away. The size of the labeled DNA fragment in the remaining hybrid is determined by electrophoresis on a denaturing acrylamide gel using appropriate size markers followed by autoradiography. The length of the fragment is a measure of the distance between the labeled end of the probe and the terminus or splice junction of the RNA. If the hybridization has been carried out with an excess of DNA probe, the intensity of the band will be proportional to the concentration of the hybridizing RNA.

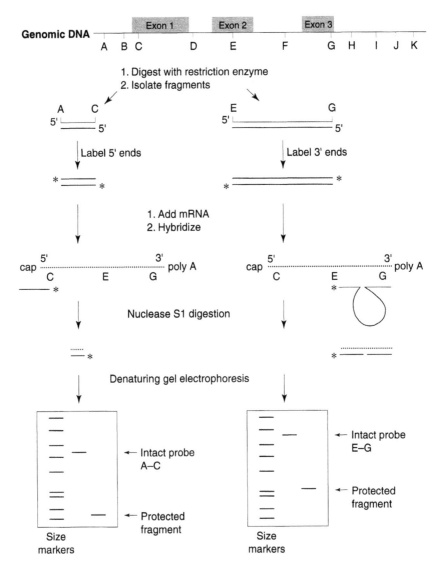

Figure 15.14. RNA mapping - Nuclease S1 protection with end-labeled nucleic acids. Mapping the position of the cap site using a probe labeled at the 5′ end and the position of an intron/exon boundary using a probe labeled at the 3′ end. Genomic fragments of DNA end-labeled with ^{32}P are hybridized to mRNA in solution and treated with nuclease S1 to remove single-stranded regions. The products are resolved on a denaturing polyacrylamide gel. The gel is dried and autoradiographed. By correlating the length of fragment protected by the 5′ labeled-probe with the DNA sequence, the position of the cap site can be deduced. By correlating the length of fragment protected by the 3′ labeled-probe with the DNA sequence, the position of the intron-exon junction can be deduced. Note that the distance of the intron/exon junction from restriction site G can not be determined using this probe, because the fragment obtained after nuclease S1 digestion is unlabeled.

There are two main problems associated with nuclease S1 mapping. First, multiple, closely spaced bands may be obtained on the gel. These may arise if several RNAs with different termini are present, but are more likely to be caused by 'nibbling' of the end of the protected fragment by nuclease S1. This is most common at A-T rich regions that frequently occur near the polyadenylation site of mRNAs. Attempts to suppress the effect are not always successful and the position of the terminus can only be defined to within a few nucleotides. 'Nibbling' is one of the main limitations of the technique. Second, the enzyme can cleave a nucleic acid duplex at internal A-T rich regions causing the appearance of spurious small bands in the gel/autoradiograph [23].

The uncertainty caused by these problems can be overcome by using a different technique.

Primer extension. A very useful technique for complementing nuclease S1 analyses is the primer extension assay. It is more sensitive than nuclease S1 mapping, gives cleaner results and is subject to different limitations.

The basis of the technique is shown in *Figure 15.15*. A 5′ end-labeled oligonucleotide or restriction fragment is annealed to the transcript and extended to the end of the RNA by reverse transcription in the presence of unlabeled nucleotides. The hybrid is denatured and the size of the denatured labeled product is determined by electrophoresis on an acrylamide gel. The size of the product reflects the distance from the label to the terminus of the RNA. An aliquot of the labeled product can be sequenced directly.

Sometimes more than one band is obtained on the gel. These arise if there is genuine heterogeneity of 5′ ends of transcripts in the cell or if the transcripts being analyzed derive from different members of a multigene family. Bands that are one or two nucleotides shorter than the full transcript length may also be generated if the reverse transcriptase has difficulty transcribing past the methylated nucleotide at the 'cap' site. Another reason for multiple bands is that the reverse transcriptase may fail to extend through regions of secondary structure in the template.

Mapping is usually carried out by more than one of the above techniques. If the results agree, this gives greater confidence that the assignment of the terminus is correct.

15.9.3 *Quantitation of transcripts*

Gene activity in cells can be monitored by measuring the steady state levels of RNA transcripts. Northern and RNA dot blots are often used in

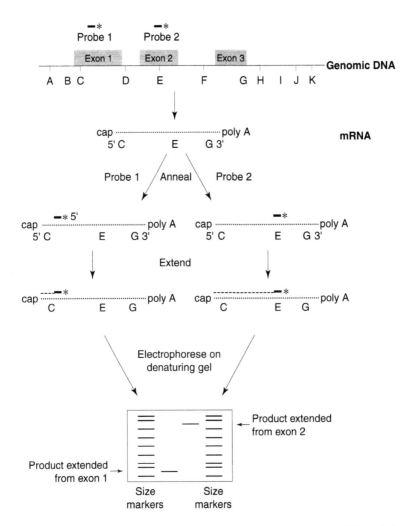

Figure 15.15. RNA mapping – Primer extension. Mapping the position of the cap site using two different oligonucleotide primers labeled at the 5′ end with ^{32}P. The mRNA is annealed in solution to the primers – in separate reactions. The primers are extend by reverse transcriptase and the products separated on a denaturing gel and autoradiographed. By correlating the length of the extension products with the DNA sequence, the transcription initiation site can be determined.

quantitative studies. They suffer several disadvantages. They require a high quality of RNA which may be difficult to achieve. In addition, the detection is relatively insensitive.

In primer extension, nuclease S1 and RNase protection an excess of probe is used in order to drive the complementary sequences into hybrid. Under these conditions, the amount of product is proportional to the amount of mRNA present. Hybridization takes place in solution which is quicker than on filters and can go to completion. So these techniques can be used as sensitive alternatives to Northern and RNA dot blot hybridization to quantitate RNA. RNase protection is particularly sensitive and is capable of detecting as little as 0.1 pg RNA. One reason for the high sensitivity is that uniform labeling allows very high specific activity of the probe. Nuclease S1 and RNase protection incorporate degradation of nucleic acid into the experimental procedure, so the quality of RNA does not need to be so high as in Northern and dot blots.

15.9.4 *Sequences expressed in one cell type and not another*

There are two hybridization-based techniques that can be used to identify sequences that are expressed in one type of cell, but not in another closely related type. Related cell types might be, for example, plant leaves that had or had not been exposed to a light stimulus, or cells cultured in the presence or absence of a hormone.

The two techniques are differential screening and subtractive hybridization. They differ fundamentally in their approach. Differential screening identifies sequences that differ between two cell types. Subtractive hybridization removes sequences that are common to both cell types and allows a cDNA of much lower complexity to be made.

Differential screening. In differential screening, a cDNA library from the light-induced cells is transformed into bacteria and plated out (*Figure 15.16*). Duplicate colony lifts are taken. One filter is probed with labeled cDNA from the light-induced cells, while the second filter is probed with labeled cDNA from uninduced cells. Sequences that are present in both cell types will give a hybridization signal on both filters. A clone derived from sequences present only in the light-induced cells will hybridize with probe only from the same cells.

Differential screening can also be used to identify sequences that are expressed at very different levels in the two cell types. The DNA on the filters is in excess and the intensity of hybridization signals on an autoradiograph varies over several orders of magnitude reflecting the abundance of particular sequences in the probe, i.e. in the original mRNA population.

Subtractive hybridization. The RNAs of interest within a cell are often expressed at low levels. This makes them difficult to detect by filter hybridization. Subtractive hybridization was developed as a technique for enriching sequences that are present at very low copy number by removing other sequences that are far more abundant [24] (*Figure 15.17*).

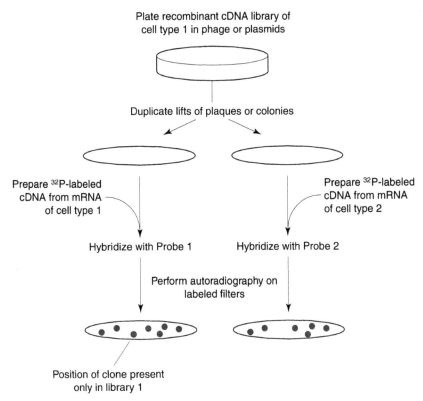

Figure 15.16. Differential screening. Duplicate plaque/colony lifts are made from a cDNA library. One filter is screened with labeled cDNA from the same cell type. The other is screened with a cDNA made from a closely-related cell type. Comparison of the two autoradiographs identifies cDNAs present in one population and absent in the other. Plaques/colonies present in only one population are picked, replated at low density and re-probed with the two probes.

Labeled cDNA made from polyadenylated mRNA of one cell type is exhaustively hybridized in solution to a 10- to 30-fold molar excess of poly(A)$^+$ (driver RNA) from a cell type that does not express the sequence of interest. Sequences that are present in both cell types form hybrids which can be removed (subtracted) by standard techniques. If the reporter molecule is biotin, for example, the hybrids can easily be removed by treatment with streptavidin-coated paramagnetic beads. A second round of hybridization is carried out with the fraction that has remained single-stranded and duplexes are again removed. The remaining single-stranded cDNA represents sequences that were originally present at low abundance and are now enriched. They can be used as a hybridization probe or can be made double-stranded and cloned. This technique can be used to identify sequences that comprise 0.01–0.05% of sequences present.

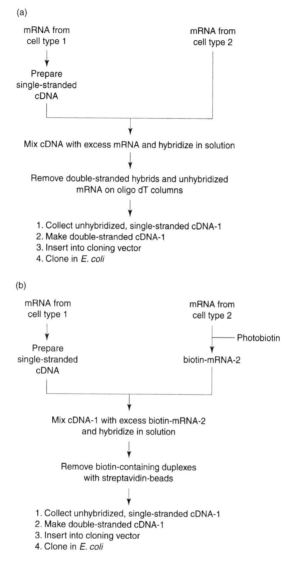

Figure 15.17. Subtractive hybridization. (a) cDNA from one type of cell is hybridized to an excess of mRNA from a closely-related cell type. Hybrids and unhybridized mRNA are removed by oligo dT chromatography. cDNA which does not hybridize is enriched in sequences that are absent from the other cell type and can be made double-stranded and cloned. The enriched cDNA usually undergoes a second round of hybridization and isolation before cloning. (b) cDNA from one type of cell is hybridized to an excess of biotinylated mRNA from a closely-related cell type. Hybrids and unhybridized biotinylated DNA are removed leaving mRNA that is enriched in cell-specific sequences.

Subtractive hybridization has one big advantage over differential screening in that far fewer clones need to be screened. However, there are several problems that limit the usefulness of the technique:

- The procedure is technically demanding.
- If the amount of available mRNA in the liquid hybridizations is not sufficiently high to drive the reaction to completion, the desired enrichment of cDNA will not be achieved.
- Separation of single- and double-stranded nucleic acids must be complete and losses at each stage of the procedure must be minimized otherwise there will be insufficient subtracted cDNA remaining after the second round of hybridization.

Hybridization techniques for detecting rare sequences are being superseded by PCR techniques because the latter are better suited for detecting rare sequences.

15.10 Hybrid selection of mRNA and hybrid arrest translation

Several years ago these two techniques were among the main ways of identifying the protein encoded by a recombinant DNA clone [25]. They have largely been superseded by PCR-based methods which are quicker, more sensitive and easier to perform. However, for completeness, the techniques are described here.

In hybrid selection, recombinant DNA, immobilized on a filter, is hybridized to mRNA. Nonspecifically bound RNA is washed away. Firmly bound mRNA is eluted and added as template to an *in vitro* protein-synthesizing system which includes a radioactive amino acid. If the DNA has selected an mRNA, a radioactive protein will be synthesized which can be detected after polyacrylamide gel electrophoresis. Hybrid selection can be used for screening large numbers of recombinant clones or for demonstrating unambiguously that a clone encodes a particular polypeptide.

In hybrid arrest translation, mRNA is hybridized to recombinant cDNA in solution. mRNA that has hybridized will be unavailable for translation when added to a protein synthesizing system *in vitro*. The products are analyzed by gel electrophoresis and compared with the products of the translation of the entire mRNA population. The two should differ by the polypeptide encoded by the complementary RNA. When the mRNA:DNA is dissociated and added to the translation system, this polypeptide should now be synthesized.

These techniques are not easy and require a fair degree of technical competence. The hybridization of complementary mRNA to the filter-bound DNA must be maximized so that there is enough mRNA to be translated in the hybrid selection procedure. It must also be maximized for hybrid arrest translation or sufficient mRNA may remain in solution to allow translation of the polypeptide. Maximising hybridization requires that the amount of DNA on the filter is high, but the following must be remembered:

- Not all DNA bound to the filter is available for hybridization probably because of steric hindrance.
- If the source of DNA is a recombinant plasmid or phage, it is best to remove the vector sequences before binding the DNA to the filter. If linearized recombinant plasmid or phage is bound to the filter, the cloned complementary sequence will represent only a fraction of the bound DNA. For example, if a 4.7 kb recombinant plasmid having a 1 kb insert is used then a maximum of 21% of the bound DNA can be complementary to the mRNA. If the vector sequences are removed, all the bound sequences will be available to hybridize to the mRNA.
- Filter hybridizations tend not to go to completion so sufficient DNA must be present to ensure that in the time available for hybridization, enough mRNA is hybridized to give a signal on subsequent translation.

  In practice, the amount of DNA to bind to the filter depends on the abundance of the mRNA of interest in the sample [26].

Other precautions should be taken when designing hybrid arrest or hybrid selection experiments. DNA should be bound to nylon filters in preference to nitrocellulose. This is in part because nylon has a better binding capacity and in part because DNA can be attached covalently. This is necessary to prevent losses of hybrids during hybridization and to prevent contamination of mRNA during elution which could inhibit translation.

It is most important to minimize RNA degradation. Hybridization is carried out in the presence of formamide to lower the T_m. This allows low incubation temperatures which is beneficial as RNA breakdown is less than at high temperatures. However, one of the main problems in these procedures is that contaminants in the formamide which accumulate on storage cause RNA degradation. It is, therefore, necessary to deionize the formamide.

References

1. **Beltz, G.A., Jacobs, K.A., Eickbush, T.H., Cherbas, P.T. and Kafatos, F.C.** (1983) *Methods Enzymol.* **100**: 266–285.
2. **Jereys, A.J., Wilson, V. and Thein, S.L.** (1985) *Nature* **314**: 67–73.

3. **Jereys, A.J., Wilson,V. and Thein, S.L.** (1985) *Nature* **316:** 76–79.
4. **Edwards, A., Civitello, A., Hammond, H.A. and Caskey, C.T.** (1991) *Am. J. Hum. Genet.* **49:** 746–756.
5. **Gill, P., Ivanov, P.L. Kimpton, C., Piercy, R., Benson, N., Tully, G., Evett, I., Hagelberg, E. and Sullivan, K.** (1994) *Nature Genet.* **6:** 130–135.
6. **Kidd, V.J., Wallace, R.B., Itakura, K. and Woo, S.L.C.** (1983) *Nature* **304:** 230–234.
7. **Conner, B.J., Reyes, A.A., Morin, C., Itakura, K., Teplitz, R.L. and Wallace, R.B.** (1983) *Proc. Natl Acad. Sci. USA* **80:** 278–282.
8. **Saiki, R.K., Bugawan, T.L., Horn, G.T., Mullis, K.B. and Erlich, H.A.** (1986) *Nature* **324:** 163–166.
9. **Dattagupta, N., Rae, P.M.M., Huguenel, E.D., Carlson, E., Lyga, A., Shapiro, J.A. and Albarella, J.P.** (1989) *Anal. Biochem.* **177:** 85–89.
10. **Serre, J.L., Taillandier, A., Mornet, E., Simon-Bouy, B., Boue, J. and Boue, A.** (1991) *Genomics* **11:** 1149–1151.
11. **Landegren, U., Kaiser, R., Sanders, J. and Hood, L.** (1988) *Science* **241:** 1077–1080.
12. **Wu, D.Y. and Wallace, R.B.** (1989) *Genomics* **4:** 560–569.
13. **Barany, F.** (1991) *Proc. Natl Acad. Sci. USA* **88:** 189–193.
14. **Sealey, P.G., Whittaker, P.A. and Southern, E.M.** (1985) *Nucleic Acids Res.* **13:** 1905–1922.
15. **Witt, M. and White, R.L.** (1985) *Proc. Natl Acad Sci. USA* **82:** 6206–6210.
16. **Nisson, P.E., Watkins, P.C., Menninger, J.C. and Ward, D.C.** (1991) *BRL Focus* **13:** 42.
17. **Flavell, R.A., Birfelder, E.J., Sanders, J.P. and Borat, P.** (1974) *Eur. J. Biochem.* **47:** 535–543.
18. **Lasky, L.A., Lev, Z., Xin, J.-H., Britten, R.J. and Davidson, E.H.** (1980) *Proc. Natl Acad. Sci. USA* **77:** 5317–5321.
19. **Anderson, M.L.M. and Young, B.D.** (1985) In: *Nucleic Acid Hybridization: A Practical Approach* (eds B.D. Hames and S.J. Higgins). IRL Press, Oxford, pp. 73–111.
20. **Sambrook, J., Fritsch, E.F. and Maniatis, T.** (1989) In: *Molecular Cloning: A Laboratory Manual, 2nd Edn.* Cold Spring Harbor Laboratory Press, Cold Spring Harbor, New York.
21. **Hames B.D. and Higgins, S.J.** (eds). (1995) *Gene Probes 2 A Practical Approach.* IRL Press, Oxford.
22. **Williams, J.G. and Mason, P.J.** (1985) In: *Nucleic Acid Hybridization: A Practical Approach* (eds Hames, B.D. and Higgins, S.J.). IRL Press, Oxford, pp. 139–160.
23. **Miller, K.G. and Solner-Webb, B.** (1981) *Cell* **27:** 165–174.
24. **Sagerstrom, C.G., Sun, B.I. and Sive, H.L.** (1997) *Annu. Rev. Biochem.* **66:** 751–783.
25. **Paterson, B.M., Roberts, B.E. and Ku, E.L.** (1977) *Proc. Natl Acad Sci. USA* **74:** 4370–4374.
26. **Mason, P.J. and Williams, J.G.** (1985) In: *Nucleic Acid Hybridization: A Practical Approach* (eds Hames, B.D. and Higgins, S.J.). IRL Press, Oxford, pp. 113–137.

16 Current trends

16.1 DNA arrays and chips

New, powerful methods have been developed that can hybridize many sequences in parallel. They involve preparation of high density arrays of sequences that are probed simultaneously. DNAs are attached to a solid surface in an ordered array so that the position or address of each DNA is known. The array is hybridized with a labeled probe and the hybrids are scanned and quantitated automatically.

There are two types of array. In the first, DNA or cDNA sequences are attached and in the second oligonucleotides are attached to the surface. The term 'DNA chip' was originally used to describe oligonucleotide arrays only, but is now also used for (c)DNA arrays.

16.1.1 Preparation of DNA arrays

The source of DNA is typically purified cDNAs that have been amplified by PCR in individual wells of a 96-well microtiter plate [1–3]. The cDNAs might be randomly picked clones from a library of interest or perhaps partially sequenced cDNAs known as expression sequence tags. Samples of about 0.05 µl are transferred by a computer-controlled printhead to predetermined positions on a series of microscope slides. These have been coated with positively charged poly-L-lysine to enhance adhesion. The density of array is about 1000 cDNAs cm^{-2}, so a microscope slide with an area of 10 cm^2 contains 10 000 hybridization targets.

16.1.2 Preparation of oligonucleotide arrays

Oligonucleotides of any sequence can be used. For example, an array may include all possible decamers or overlapping oligomers containing all known sequence changes in a gene implicated in disease, such as the *ras* gene which is mutated in many cancers [4,5].

Oligonucleotides of specified sequence can either be synthesized *in situ* or prepared separately and attached postsynthetically. A variety of surfaces such as microscope slides, polypropylene films and optical fibers can be used (ref. 6 and references therein).

For synthesis *in situ*, oligonucleotide synthesis is directed to the appropriate address using masks that minimize the number of condensation steps necessary [5]. Use of photolabile protection groups and ink-jet technology have speeded up the process. The maximum length attainable for oligonucleotides in an array is probably limited by the yield at each step, but currently arrays of 25-mers are achievable and are probably sufficiently long for most purposes. A density of oligonucleotides of at least 400 000 1.6 cm^{-2} can be achieved at present.

Not all laboratories have the capability to prepare arrays. However, they are available commercially (Affymetrix) together with flow cells in which the hybridization takes place.

Label. Fluorescent labels are popular because they can be detected and quantitated automatically. Different probes can be labeled with different fluorophors, then mixed together and used to probe an array in a single hybridization reaction. This avoids the problem of slight variations in hybridization efficiency if the probes are analyzed in separate reactions.

Biotin can be attached to probes and subsequently detected in hybrids by binding to streptavidin conjugated to a dye such as phycoerythrin and lazer light illumination [7]. Particles such as silicon can be attached to probes and detected by light scattering [8].

Applications. The principal applications of arrays are in genomics, diagnostics, gene expression analysis and DNA sequencing.

16.1.3 Gene expression analyses

Any given cell expresses only a limited number of its encoded genes. The number of genes expressed and the level of expression varies according to factors such as cell type, stage of development and response to external stimuli. Understanding gene expression and how it is regulated depends on analyzing many transcripts simultaneously. Hybridization with labeled RNAs to targets on chips allows quantitative and qualitative measurement of transcripts to be made with a sensitivity of one to five transcripts per cell [9]. Transcripts from two different cell types can be labeled with different fluorescent dyes. The probes are mixed and hybridized simultaneously to chip-bound targets. The two-color fluorescence of hybrids allows differential gene expression to be monitored [2]. Use of alternative splice sites can also be detected.

Detection of mutations and polymorphisms. Methods are currently being developed for screening patient and carrier populations for mutations [10].

Arrays can be prepared to include oligonucleotides that cover known mutations within a gene. An improved method is to include in the array of oligonucleotides sequences that have the other three bases at the sites of known mutations. The alternative bases are the center of the oligonucleotides where mismatching has the most disruptive effect on the stability of hybrids. On hybridization, the mutant gene should hybridize to one of these mutant sequences. This not only identifies the mutant base but gives added assurance that the hybridization is working.

Test arrays of 100 000 oligonucleotides of 20-mers have been used to screen for mutations in exon 11 of the human *BRCA*1 gene [11]. This exon codes for about 60% of the protein. Sequences covered the normal sequence as well as oligonucleotides covering insertions, deletions and substitutions. All known mutations were detected with no false positives. Diagnostic arrays are likely to be of increasing importance and already arrays of HIV oligonucleotides are available.

In several conditions, such as the thalassemias and hemoglobinopathies, mutations in several different positions within the gene can give rise to disease. Overlapping oligonucleotide arrays are potentially well-suited to screening rapidly for the presence of mutations. Arrays of overlapping oligonucleotides of the gene to be screened are hybridized to DNA probes from normal and patient samples. By labeling the probes from each source with a different fluor, hybridization can be carried out with both probes at once. Automatic scanning of the array makes diagnosis objective.

16.1.4 DNA sequencing by hybridization

As noted above, an array can contain all possible oligonucleotides of a chosen length, n. Hybridizing a probe to such an array provides information on all the constituent sequences of length n in the probe [5,10]. Sufficient information can be generated to allow assembly into a reconstructed sequence of the probe. This approach is known as sequencing by hybridization (*Figure 16.1*).

When the probe is hybridized to the array, some oligonucleotides will hybridize and some will not. The sequence of those oligonucleotides that have hybridized can be aligned by overlapping their sequences to

```
5' GGATGAACTG 3'
   GATGAACTGT
    ATGAACTGTT
     TGAACTGTTA
      GAACTGTTAC
```

Figure 16.1. Solid phase sequencing. Sequencing of a hypothetical stretch of DNA, 5'- - -GTAACAGTTCATCC- - -3'. The DNA is labeled with a fluorescent label and hybridized to a DNA chip containing all possible decamers. The sequences of all the 10-mers which hybridize to the DNA are noted and aligned by computer into a series of overlapping sequences such that one 10-mer overlaps the preceding and succeeding 10-mer by 9 nucleotides. From the alignment of the oligomers above, the complementary sequence of the DNA is: 5'- - -GGATGAACTGTTAC- -3'.

reconstruct the complete sequence. Each nonterminal n-mer will overlap the next by $n-1$ nucleotides. This method of sequencing is useful in terms of speed, cost, ease of automation and quality of data.

However, there are two main problems with the technique. First, it is a major operation to align overlapping sequences correctly. Second, the technique runs into difficulties if the probe contains repeated sequences because the overlaps will not be unique. The reconstruction stops at a repeated sequence. For these reasons, sequencing by hybridization is usually carried out as a resequencing operation for confirming a known sequence.

Confirmation of the sequence of the entire human mitochondrial genome has been obtained using oligonucleotide arrays [12]. This provides a useful basis for applications such as analysis of mitochondrial DNA from different species or screening DNA from patients with disorders that are thought to be associated with changes in mitochondrial DNA sequence.

16.2 Probes

16.2.1 Developments in hybridization with oligonucleotide and peptide nucleic acid probes

In the last few years there has been much interest in developing hybridization with other types of oligonucleotides and this has been fueled by their potential use as antisense and antigene therapeutic agents. An antisense agent binds to mRNA and prevents its translation.

The aim of antigene agents is to correct a mutation in the DNA of a cell by gene conversion. Oligoribonucleic acid, chimeric oligonucleotides containing both deoxyribonucleotide and ribonucleotide residues are potentially useful.

Peptide nucleic acids (PNAs) are also being developed. These molecules mimic nucleic acid. They have a pseudo-peptide backbone of *N*-(2-aminoethyl)-glycine units to which bases are attached. The backbone is nonionic and achiral. The spacing of the glycine residues is optimal for hybridization to nucleic acid. Peptide nucleic acids can form hybrids with RNA, DNA and PNAs. The hybridization properties of PNAs and the increased stability of PNA:DNA hybrids over the corresponding DNA:DNA hybrids under physiological conditions suggests that they may be useful as antigene agents. The properties of PNA:RNA hybrids suggest that they may be useful as antisense agents [13]. PNA:DNA hybrids are more destabilized by a mismatch than the corresponding DNA:DNA hybrid so they may become useful probes for the presence of mutations. PNAs do not serve as primers for PCR [14].

16.2.2 *Molecular beacons*

New applications of hybridization promise to be opened up by the development of probes termed molecular beacons. These are single-stranded oligonucleotides that fluoresce when they hybridize [15]. The probes have a stem–loop structure (*Figure 16.2*). The stem is composed of complementary sequences that flank the probe sequence in the loop. The sequences of the stem and loop bear no resemblance to each other. A fluorophore is bound to one end of the stem and a group that quenches fluorescence is attached to the other. The unhybridized molecule does

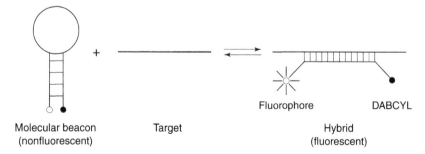

Molecular beacon   Target   Hybrid
(nonfluorescent)   (fluorescent)

Figure 16.2. Principle of detection of hybrids with molecular beacons. When unhybridized, the hairpin stem of the beacon keeps the quenching group close to the fluorophore and prevents fluorescence. When the beacon hybridizes the two ends of the probe are too far apart for the quenching group to have any affect on the fluorophore and the hybrid fluoresces.

not fluoresce because the base-pairing in the stem keeps these groups close to each other so that the fluorescence of the fluorophore is quenched by the quenching group. When the beacon hybridizes, a conformational change occurs which disrupts the base-pairing in the stem and as a consequence the quenching group is too far away to quench the fluorophore. Hence, the hybrid fluoresces.

These molecules can discriminate between targets that differ by a single base [16]. Molecular beacons have some useful properties. It is not necessary to remove nonhybridizing molecules to detect hybrids so this saves time. These molecules are potentially useful for introducing into cells and following the synthesis of RNA in real time. By attaching different fluorophores to different probes, many different sequences can be detected simultaneously.

References

1. **Schena, M., Shalon, D., Davis, R.W. and Brown, P.O.** (1995) *Science* **270:** 467–470.
2. **Shena, M.** (1996) *BioEssays* **18:** 427–431.
3. **Adams, M.D.** (1996) *BioEssays* **18:** 261–262.
4. **Southern, E.M., Maskos, U. and Elder, J.K.** (1992) *Genomics* **13:** 1008–1017.
5. **Southern, E.M.** (1996) *Trends Genet.* **12:** 110–115.
6. **Ferguson, J.A., Boles, T.C., Adams, C.P. and Wall, D.R.** (1996) *Nature Biotechnol.* **14:** 1681–1684.
7. **Lockhardt, D.J., Dong, H., Byrne, M.C., Follettie, T., Gallo, M.V** *et al.* *Nature Biotechnol.* **14:** 1675–1680.
8. **Stimpson, D.I., Hoijer, J.V., Hsieh, W.T., Jou, C. Gordon, J, Theriault, T.** *et al.* (1995) *Proc. Natl Acad. Sci. USA* **92:** 6379–6383.
9. **de Saizieu, A., Certa, U., Warrington, J., Gray, C., Keck, W. and Mous, J.** (1998) *Nature Biotechnol.* **16:** 45–48.
10. **Wallace, R.W.** (1997) *Molec. Med. Today* **3:** 384–389.
11. **Hacia, J.G., Brody, L.C., Chee, M.S., Fodor, S.P.A. and Collins, F.S.** (1996) *Nature Genet.* **14:** 441–447.
12. **Chee, M., Yang, R., Hubbell, E., Berno, A., Huang, X.C. Stern, D.** *et al.* (1996) *Science* **274:** 610–613.
13. **Jensen, K.K., Ørum, H., Nielsen, P.E. and Norden, B.** (1997) *Biochemistry* **36:** 5072–5077.
14. **Ørum, H., Nielsen, P.E., Egholm, M., Berg, R.H., Buchardt, O. and Stanley, C.** (1993) *Nucleic Acids Res.* **21:** 5332–5336.
15. **Tyagi, S. and Kramer, F.R.** (1996) *Nature Biotechnol.* **14:** 303–308.
16. **Tyagi, S., Bratu, D.P. and Kramer, F.R.** (1998) *Nature Biotechnol.* **16:** 49–53.

Appendix A

Glossary

Allele: one of two or more different forms of a gene at a particular locus.

Anneal: the (re-) establishment of base pairing between complementary strands of DNA, RNA or a DNA and an RNA strand.

Antisense strand: a DNA strand that has a sequence complementary to mRNA.

Autoradiography: the detection of radioactively labeled molecules present in, for example, a filter or gel by exposing to X-ray film.

Background: the nonspecific binding of probe to the filter. Background is generally lowered by prehybridizing in solutions that precoat the filter and washing the filter in low ionic strength solution ($0.1 \times$ SSC) at high temperature ($60°C$ or above).

cDNA: a single-stranded DNA sequence complementary to RNA and synthesized from it by reverse transcription.

Complexity: the total length of nucleic acid that could be created by ligating together all the different sequences within a cell/ organism. Complexity takes no account of how many times a particular sequence is repeated.

$C_o t$: the controlling factor for estimating the completion of a DNA:DNA hybridization/reassociation reaction. It is the product of DNA concentration at zero time of incubation (in mol nucleotide per litre) and time of incubation (in seconds).

$C_o t_{1/2}$: the $C_o t$ value at which 50% of the DNA sequences have reassociated.

Criterion: the difference in temperature between the incubation temperature, T_i, and the melting temperature, T_m, of the duplex.

Cross-hybridization: the binding of a single-stranded nucleic acid to a target that is not its perfect complement. This may be allowed by using 'relaxed' hybridization conditions.

Denaturation or melting: breaking the weak bonds that hold complementary strands of double-stranded DNA, RNA or DNA:RNA together so that the strands dissociate. The term is also used for destroying secondary structure in single-stranded nucleic acids.

Homology: the degree of identity between nucleotide sequences of two related, but not fully complementary DNA or RNA molecules. Thus, 80% homology means that on average the sequence of 80 of 100 nucleotides is identical.

Hybridization: the pairing of complementary RNA and DNA strands to give an RNA:DNA hybrid. It is also used to describe the pairing of two single-stranded DNA or RNA strands.

Multigene family: a set of identical or related genes present in the same organism usually coding for a family of proteins that have a similar primary structure.

Northern blotting: a procedure for transferring denatured RNA from an agarose gel to an inert surface such as a nitrocellulose filter for detection of complementary sequences by molecular hybridization.

PCR (polymerase chain reaction): an *in vitro* method for amplifying a stretch of DNA.

Polymorphism: a variation in DNA sequence that occurs at high frequency in a population.

Probe: a labeled DNA, RNA or oligonucleotide molecule used to detect a complementary strand by molecular hybridization.

Pseudo-first order reaction: if one of the nucleic acid species in a hybridization reaction is in vast excess over the other, the reaction follows first order kinetics. The concentration of the species in excess is hardly depleted by duplex formation.

Relaxed conditions: hybridization conditions are said to be relaxed if they permit duplexes to form that have some mismatched bases in the complementary strands (the hybridizing sequences are related, but not identical). High salt concentration, presence of formamide and low temperatures during hybridization allow formation of mismatched duplexes. High salt concentration and low temperatures of washing allow imperfectly matched hybrids to persist.

Renaturation: the re-establishment of a DNA double helix or intrastrand hairpin structures in RNA after denaturation.

RFLP (restriction fragment length polymorphism): the variation in size of fragment that occurs when DNA from two or more individuals are digested with the same restriction endonuclease. The variation may arise from a deletion, an insertion or a change of sequence in the enzyme recognition site.

Second order reaction: if the concentration of the complementary strands in a reassociation/hybridization reaction is equal, the reaction follows second-order kinetics.

Southern blotting: a capillary blotting procedure for transferring denatured DNA from a gel to an inert surface such as a nitrocellulose filter for detection of complementary sequences by molecular hybridization.

Specificity (of hybridization): the extent to which perfectly matched hybrids are formed in preference to hybrids containing mismatches. If the stringency of hybridization is high, the specificity will be high and only perfectly matched duplexes will form.

Stringency: the combination of temperature, salt and formamide concentration that determines the specificity of hybridization. Conditions are said to be stringent if they allow only well-matched hybrids to form and relaxed if they allow imperfectly matched hybrids to form.

Target: a nucleic acid molecule which is to be detected by hybridization or amplified by PCR.

T_m **(melting temperature):** for DNA, the temperature at which 50% of bases have melted. For hybrids containing oligonucleotides, the temperature at which 50% of the molecules have dissociated.

Appendix B

Solution hybridization equations

Reassociation of DNA depends on the random collision of single strands. When native DNA is denatured, the concentration of each strand equals that of the complement and the rate of disappearance of single strands follows a second-order reaction according to the equation:

$$\frac{-\mathrm{d}C}{\mathrm{d}t} = kC^2 \tag{B.1}$$

where C is the concentration of nucleotides in each of the nucleic acid strands (moles per litre), t is the time in seconds and k is the rate constant ($1\ \mathrm{s}^{-1}\ \mathrm{mol}$).

The equation can be rearranged and integrated to give:

$$\int \frac{\mathrm{d}C}{C^2} = -\int k\mathrm{d}t \tag{B.2}$$

$$-C^{-1} = -kt + \text{constant} \tag{B.3}$$

The integration constant can be calculated from the initial conditions. When $t = 0$, $C = C_0$, so $-C_0^{-1} = \text{constant}$.

So substituting in Equation B.3

$$\frac{1}{C} - \frac{1}{C_o} = kt \tag{B.4}$$

By multiplying both sides by C_o and adding 1 to both sides:

$$\frac{C_o}{C} - \frac{C_o}{C_o} + 1 = kC_o t + 1 \tag{B.5}$$

and taking the reciprocal,

$$\frac{C}{C_o} = \frac{1}{1 + kC_o t} \tag{B.6}$$

At $t_{1/2}$, $C = 0.5C_o$ and $C/C_o = 0.5$

Equation B.6 becomes

$$\frac{1}{1 + kC_o t_{1/2}} = 0.5 \tag{B.7}$$

$$or \ k \ C_o t_{1/2} = 1$$

Thus

$$C_o t_{1/2} = \frac{1}{k} \ or \ t_{1/2} = \frac{1}{kC_o} \tag{B.8}$$

So the controlling factor for the reassociation is the product of the DNA concentration (C_o) and the time of reaction (t).

DNA:RNA hybridization

DNA in excess

When DNA is in large excess, with the RNA present in minute quantities as a radioactive tracer, two competing reactions occur.

$$\text{DNA} + \text{DNA} \rightarrow \text{DNA:DNA}$$
$$\text{DNA} + \text{RNA} \rightarrow \text{DNA:RNA}$$

in which k is the second order rate constant ($1 \ \text{mol}^{-1}$ s) that varies with the conditions of the reaction. Assume that the rate constant is the same for both reactions (this is only approximate in practice). The rate of disappearance of single-stranded DNA is given by

$$\frac{-\text{d}C}{\text{d}t} = kC^2 \tag{B.9}$$

where C is the concentration of single-stranded DNA (moles nucleotides per litre) and t is the time in seconds. This can be integrated from initial conditions $t = 0$ and $C = C_0$ to give:

$$\frac{C}{C_o} = \frac{1}{1 + kC_o t} \tag{B.10}$$

or

$$C = \frac{C_o}{1 + kC_o t} \tag{B.11}$$

The rate of disappearance of single-stranded RNA is given by the equation:

$$\frac{-\mathrm{d}R}{\mathrm{d}t} = kRC \tag{B.12}$$

where R is the concentration of single-stranded RNA (mol nucleotide l^{-1}).

Substitution in Equation B.10 gives

$$\frac{-\mathrm{d}R}{\mathrm{d}t} = kR\frac{(C_0)}{(1+kC_0t)} \tag{B.13}$$

or

$$\frac{\mathrm{d}R}{R} = -kC_0\frac{(1)}{(1+kC_0t)}\,\mathrm{d}t \tag{B.14}$$

This can be integrated from initial conditions of $t = 0$ and $R = R_0$ to give:

$$\ln\frac{R}{R_0} = \ln\frac{1}{1+kC_0t} \tag{B.15}$$

or

$$\frac{R}{R_0} = \frac{1}{1+kC_0t} \tag{B.16}$$

So the RNA sequences in a DNA excess reaction hybridize at the same rate as the DNA sequences. Since the rate depends only on the initial DNA concentration, C_0 this type of reaction is said to be DNA-driven.

Since the RNA tracer hybridizes with the same kinetics as DNA, the $C_0t_{1/2}$ for a species of RNA can be used to determine the frequency class of DNA from which it was transcribed.

It may be more convenient to study the properties of RNA through synthesizing a cDNA followed by hybridization analysis of the cDNA. For these purposes the cDNA behaves exactly like the RNA.

RNA in excess
In an RNA excess or RNA-driven reaction, the following two reactions occur:

$$\text{DNA} + \text{RNA} \rightarrow \text{DNA:RNA}$$
$$\text{DNA} + \text{DNA} \rightarrow \text{DNA:DNA}$$

Since the DNA is present only as a tracer, its concentration is very low and the DNA:DNA reaction can usually be neglected.

The rate of disappearance of single-stranded DNA is given by the equation

$$\frac{-dC}{dt} = kRC \qquad (B.17)$$

Since the RNA concentration is in vast excess over that of DNA, the RNA concentration does not change appreciably during the reaction, so $R = R_0$.

So equation B.17 becomes

$$\frac{-dC}{dt} = kR_0C \qquad (B.18)$$

or

$$\frac{-dC}{C} = -kR_0dt \qquad (B.19)$$

On integration from the initial conditions of $T = 0$ and $C = C_0$

$$\ln\frac{C}{C_0} = -kR_0t \qquad (B.20)$$

or

$$\frac{C}{C_0} = e^{-kR_0t} \qquad (B.21)$$

So an RNA-driven reaction is a pseudo-first order reaction. The value of R_0t when the reaction is half complete ($R_0t_{1/2}$) is given by:

$$R_0t_{1/2} = \frac{ln2}{k} \qquad (B.22)$$

The complexity of the RNA population can be derived from either the $R_0t_{1/2}$ or the endpoint of an RNA-driven reaction.

Appendix C

Useful information

Sources to check for oligonucleotide design and melting temperatures

Many software packages are available for designing an oligonucleotide probe, checking the proposed probe against sequence databases and for determining the T_m of hybrids. It is beyond the scope of this book to list them all. Excellent discussions of the hardware and software available and the databases containing probe information can be found in [1,2]. A programme, OLIGO, for selecting an optimal oligonucleotide probe and determining the T_m is described in [3]. Some software is available on the Internet at low or nominal cost.

Useful programmes include:
T_m determination:
  http://alces.med.umn.edu/rawtm.html
Pedro's tools:
  http://www/ nwfcs.noaa.gov/protocols/pedro/rt-1.html
Oligonucleotide determination:
  http://www/ nwfcs.noaa.gov/protocols/oligoTMcalc.html
The PRIME programme of the Genetics Computer Group Inc's Wisconsin sequence analysis package.

References

1. **Fuchs, R. and Cameron, G.N.** (1995) In: *DNA Cloning 3 – Complex Genomes: A Practical Approach*. (D.M. Glover and B.D. Hames eds), IRL Press, Oxford.
2. **Woollard, P.M. and Williams G.** (1997) In: *Genome Mapping: A Practical Approach*. (P.H. Dear ed.), IRL Press, Oxford.
3. **Rychlik, W. and Rhodes, R.E.** (1989) *Nucl. Acids Res.* **17:** 8543–8551.

Conversion factors

Spectrophotometric conversions

$1 A_{260}$ unit of double-stranded DNA $\equiv 50\,\mu g\ ml^{-1}$

$1 A_{260}$ unit of single-stranded DNA $\equiv 33\,\mu g\ ml^{-1}$

$1 A_{260}$ unit of single-stranded RNA $\equiv 40\,\mu g\ ml^{-1}$

$1 A_{260}$ unit of random sequence oligonucleotides $\equiv 33\,\mu g\ ml^{-1}$

For oligonucleotides of known sequence, it is best to determine the concentration by summing the number of each nucleotide present and using the molar extinction co-efficient.

DNA molar conversions

$1\,\mu g$ double-stranded DNA of 1 kb in length $\equiv 1.52$ pmol
   (3.04 pmol ends)

$1\,\mu g$ double-stranded DNA of 2 kb in length $\equiv 0.76$ pmol
   (1.52 pmol ends)

1 pmol double-stranded DNA of 1 kb in length $\equiv 0.66\,\mu g$

1 pmol double-stranded DNA of 2 kb in length $\equiv 1.32\,\mu g$

Radioactivity

Radioactivity is expressed in terms of the becquerel (Bq).

1 Bq = 1 disintegration per second.

However, many investigators still use the older term of the Curie (Ci).

$1\ Ci = 3.7 \times 10^{10}$ Bq.

$1\,\mu Ci = 2.2 \times 10^{6}$ disintegrations per minute (d.p.m.).

Appendix D

Composition of stock solutions

$20 \times$ SSC	3 M NaCl 0.3 M trisodium citrate adjust pH to 7.0 with NaOH
$20 \times$ SSPE	3.6 M NaCl 0.2 M sodium phosphate buffer, pH 7.4 20 mM EDTA
$10 \times$ MOPS buffer	0.2 M MOPS (3-(N-morpholino)propane- sulphonic acid) 50 mM Na acetate, pH 7.0 10 mM EDTA Store in the dark
$100 \times$ Denhardt's solution	2% Ficoll (mol.wt. 400 000) 2% polyvinylpyrrolidone (400 000) 2% bovine serum albumen (Fraction V)
20% SDS	Store at room temperature
Sonicated DNA (usually calf thymus or salmon sperm DNA)	Add DNA to water to about 5 mg ml^{-1}. Stir until dissolved – it may take several hours. Sonicate or shear 10–15 times through a gauge 17 syringe needle to a length of 400–800 bp. Check the size by agarose gel electrophoresis. Adjust the concentration to 5 mg ml^{-1}. Store at $-20°$C in aliquots.

Deionized formamide (note 1)	To 200 ml formamide add 10 g mixed bed resin such as AG501-X8(D) (BioRad) or Amberlite MB3. Stir for 1 h at room temperature. Filter through Whatman No. 1 paper to remove the resin. Store in aliquots in the dark, tightly capped bottles (filled to exclude air) at 48°C.

Note

1. Formamide is a teratogen. Handle with care and wear gloves. Freshly opened bottles of reagent-grade formamide may not require deionization. However, if the color is at all yellow, the above deionization procedure should be carried out.

Appendix E

Suppliers

Affymetrix, 3380 Central Expressway, Santa Clara, CA 95051, USA.

Aldrich Chemical Company Ltd, The Old Brickyard, New Road, Gillingham, Dorset, SP8 4JL, UK.

Amersham International plc, Lincoln Place, Green End, Aylesbury, Buckinghamshire HP20 2TP, UK.

Amersham North America, 2636 South Clearbrook Drive, Arlington Heights, IL 60005, USA.

Anachem Ltd, 20 Charles Street, Luton, Bedfordshire LU2 0EB, UK.

Anderman and Co. Ltd, 145 London Road, Kingston-upon-Thames, Surrey KT17 7NH, UK.

Applied Biosystems, Division of Perkin-Elmer, 850 Lincoln Centre Drive, Foster City, CA 94404, USA.

Applied Biosystems, Division of Perkin-Elmer Ltd, Kelvin Close, Birchwood Science Park North,Warrington, Cheshire WA3 7PB, UK.

Appligene SA, Parc d'Innovation, Route du Rhin, BP72, F-767402 Illkirch, France.

Appligene, Pinetree Centre, Durham Road, Birtlay Chester-le-Street, Durham DH3 2TD, UK.

Bio-Rad Laboratories Ltd, Bio-Rad House, Maylands Avenue, Hemel Hemstead, Hertfordshire HP2 7TD, UK.

Bio-Rad Laboratories, 2000 Alfred Nobel Drive, Hercules, CA 94547, USA.

Boehringer Mannheim Corporation, Biochemical Products, 9115 Hague Road, PO Box 50414, Indianapolis, IN 46250, USA.

Boehringer Mannheim (Diagnostics and Biochemicals) Ltd, Bell Lane, Lewes, East Sussex BN17 1LG, UK.

Boehringer Mannheim GmbH, Biochemica, Sandhofer Str. 116 Postfach 310120 Mannheim 31 D-6800, Germany.

British Biotechnology Products Ltd, 4–10 The Quadrant, Barton Lane, Abingdon, Oxfordshire OX14 3YS, UK.

Cambridge Biosciences Ltd, Stourbridge Common Business Centre, Swann's Road, Cambridge CB5 8LA, UK.

Cambridge Research Biochemicals Inc., Fairfax Research Room 205, Wilmington, DE 19897, USA.

Cambridge Research Biochemicals Ltd, Gadbrook Park, Northwich, Cheshire CW9 2RA, UK.

Clontech Laboratories Inc., 1020 East Meadow Circle, Palo Alto, CA 94303-4607, USA.

Clontech Laboratories Inc., in Europe, *see* Cambridge Biosciences Ltd.

Cruachem Ltd, Todd Campus, West of Scotland Science Park, Acre Road, Glasgow G20 0UA, UK.

Dupont (UK) Ltd, NEN, New Life Sciences Products Products, Wedgewood Way, Stevenage, Hertfordshire SG1 4QN, UK.

Dupont NEN Research Products, 549 Albany Street, Boston, MA 02118, USA.

Dynal Inc., 5 Delaware Drive, Lake Sucess, NY11042, USA.

Dynal International, PO Box 158 Skøyen, N0212, Oslo, Norway.

Dynal UK Ltd, 10 Thursby Road, Croft Business Park, Broomsborough, Wirral, Merseyside L63 3PW, UK.

Eastman Kodak, Acorn Field, Knowsley Industrial Park North, Liverpool L33 72X, UK.

Eastman Kodak, Rochester, New York 14650, USA.

Fisher Scientific Co., 711 Forbes Avenue, Pittsburgh, PA 15219-4785, USA.

Genetic Research Instrumentation Ltd, Gene House, Dunmow Road, Felsted, Dunmow, Essex CM6 3LD, UK.

Gibco/BRL, PO Box 68, Grand Island, NY 14072-0068, USA. or Gibco Life Technologies, 8400 Helgerman Court, Gaithersburg, MD 20877, USA.

Gibco/BRL, for UK *see* Life Technologies Ltd.

Hybaid Ltd, 111–113 Waldegrave Road, Teddington, Middlesex TW11 8LL, UK.

Hybaid, National Labnet Corporation, PO Box 841, Woodbridge, NJ 07095, USA.

ICN Biochemicals Ltd, Unit 18, Thame Park Business Centre, Wenman Road, Thame, Oxfordshire OX9 3XA, UK.

ICN Biomedicals Inc., 3300 Hyland Avenue, Costa Mesa, CA 92626, USA.

International Biotechnologies Inc., (A Kodak Company) 36 Clifton Road, Cambridge CB1 4ZR, UK.

International Biotechnologies Inc., PO Box 9558, 25 Science Park, New Haven, CT 06535, USA.

Invitrogen Corporation: distributed in UK by British Biotechnology Products Ltd.

Invitrogen Corporation, 3985 B Sorrento Valley Boulevard, San Diego, CA 92121, USA.

Life Technologies Industrial Bioproducts, 8717 Grovemont Circle, PO Box 6009, Gaithersburg, MD 20884 USA.

Life Technologies Ltd, PO Box 35, 3 Fountain Drive, Inchinnan Business Park, Paisley PA4 9RF, UK.

Mallinckrodt Inc., 222 Red School Lane, Phillipsburg, NJ 00865, USA.

Millipore Corp., PO Box 9125, 80 Ashby Road, Bedford, MA 01730, USA.

Millipore (UK) Ltd, The Boulevard, Blackmoor Lane, Watford, Hertford-shire WD1 8YW, UK.

New England Biolabs, 32 Tozer Road, Beverley, MA 01915-5599, USA.

New England Biolabs/C.P Laboratories, PO Box 22, Bishop's Stortford, Hertfordshire CM23 3DH, UK.

Perkin-Elmer Cetus, 761 Main Avenue, Norwalk, CT 06856, USA.

Perkin-Elmer Ltd, Analytical Instruments, Chalfont Road, Seer Green, Beaconsfield, Buckinghamshire, HP9 1QA, UK.

Perkin-Elmer Holding GmbH, Bahnhofstrasse 30, D-8011 Vaterstetten, Munich, Germany.

Pharmacia LKB Biotechnology Inc., 800 Centennial Avenue, PO Box 1327, Pistcataway, NJ 08855-1327, USA.

Pharmacia LKB Biotechnology Ltd, 23 Grosvenor Road, St Albans, Hertfordshire AL1 3AW, UK.

Pierce, PO Box 117, Rockford, IL 61105, USA.

Promega Corp., 2800 Woods Hollow Road, Madison WI 53711-5399, USA.

Promega Ltd, Delta House, Enterprise Road, Chilworth Research Centre, Southampton S01 7NS, UK.

QIAGEN GmbH, Max-Volmer-Str. 4, Hilden-D-40724, Germany.

QIAGEN Inc., 9600 De Soto Avenue, Chatsworth, CA 91311, USA.

QIAGEN, distributed in UK by Hybaid.

Research Genetics Inc., 2130 Memorial Parkway, Huntesville, Alabama 35801, USA.

Sarstedt Ltd, 68 Boston Road, Beaumont Leys, Leicester LE4 1AW, UK.

Sartorius Ltd, Longmead, Blenheim Road, Epsom, Surrey KT9 9QN, UK.

Sartorius North America Inc., 140 Wilbur Place, Bohemia, Long Island, NY11706, USA.

Shleicher and Schuell Inc., distributed in UK by Anderman and Co Ltd.

Shleicher and Schuell Inc., 10 Optical Avenue, PO Box 2012, Keene, NH 03431, USA.

Sigma Chemical Company, 3050 Spruce Street, PO Box 14508, St Louis, MO 63178-9916, USA.

Sigma Chemical Company (UK), Fancy Road, Poole, Dorset BH12 4QH, UK.

Stratagene Inc., 11099 North Torrey Pines Road, La Jolla, CA 92037, USA.

Stratagene Ltd, Unit 140, Cambridge Innovation Centre, Cambridge Science Park, Milton Road, Cambridge CB4 4GF, UK.

Stratech Scientific Ltd, 61–63 Dudley Street, Luton, Bedfordshire LU2 0NP, UK.

Techne Inc., 743 Alexander Road, Princeton, NJ 08540, USA.

Techne Ltd, Hinxton Road, Duxford, Cambridge CB2 4PZ, UK.

Tropix Inc., 47 Wiggins Avenue, Bedford, MA 01730, USA.

Vector Laboratories, 16 Wulfric Square, Bretton, Peterborough PE3 8RF, UK.

Vector Laboratories, 30 Ingold Road, Burlingame, CA 94010, USA.

Whatman Laboratory Products Inc., 9 Bridewell Place, Clifton, NJ 07014, USA.

Whatman International Ltd, Whatman House, St Leonards Road, Maidstone, Kent ME16 0LS, UK.

Wheaton Science Products, 100 North 10th Street, Millville, NJ 08332, USA.

Wheaton Scientific, distributed by UK in Jencons Scientific Ltd, Cherrycourt Way Industrial Estate, Stanbridge Road, Leighton Buzzard, Bedfordshire LU7 8UA, UK.

Index

Abundance of mRNA, 43–44, 189
Accelerator hybridization, 73–74, 81, 143
Alkaline phosphatase (AP), 61, 151–155, 157–161
  labeling of probes, 74, 92
    affect on T_m of hybrids, 74
  detection of hybrids, 134–135
  chemiluminescent substrates for, 155–157
  chromogenic substrates for, 152–157
  multiprobe detection, 153–155, 158–159
Allele-specific oligonucleotide hybridization, 181–184
Alu repetitive sequences
  use as probes, 187
  in probes, 91, 186–187
  removal, suppression, 186–187
AMPPD, 61, 155–157, 159
Autoradiography
  choice of film, 148–149, 161
  definition, 211
  intensifying screen, 147–148, 149–150
  procedure, 149–151

Background
  definition, 66, 211
  high, causes of
    alkali fixation, 103, 108
    charged nylon filters, 97, 108
    dextran sulfate, 143
    excess reporter molecules, 132
    hybridization accelerators, 74, 143
    in colony/plaque lifts, 106
    in indirect detection, 152
    probe, high concentration, 143, 145
    probe, splashes, drying, 139, 145, 149, 160
    tailed probes, 128
    trapped probe, 142
    trouble shooting, 168–170
    unincorporated nucleotides, 129
    vacuum blotting, 100

    wet filter, 149
  low, causes of
    naphthol AS, 153
    nitrocellulose filters, 97–98
    radioactive probes, 92, 95
    SDS buffer, 140
  problems, 66, 68, 74, 129, 138
  reduction of, 63, 79, 129, 138, 141, 143, 160, 161, 173, 187
Base composition, effect on
  hybridization rate, 51, 68
  hybridization stability, 6–7, 53, 68
  T_m, 6–7, 30, 68–69, 77–79
Base-pairing, 1–2, 4, 79
Base stacking, 2–3, 5–6, 77
BCIP, 92–95, 116–117, 119, 122, 124, 126, 132, 152, 154, 160, 185–186, 200, 206
Biotin-(strept)avidin, 152, 154, 158, 160–161, 185–186, 200, 206
Biotinylated probe, detection of in hybrids, 154, 158
Binding nucleic acid to filter
  transfer nucleic acid to filter, 98–102
    of DNA from gels, 98–101
    of DNA/RNA to blots, 101–102
    of phage/colonies, 98
    of RNA from gels, 101
    protocols, 105–114
  immobilization of nucleic acid
    alkali fixation, 103–104
    baking, 102
    protocols, 105–114
    UV fixation, 102–103
BLOTTO, 138
  see also Nonfat milk

Calf intestinal alkaline phosphatase (CIAP), 126–127
Calf thymus phosphatase, 126
Cap site mapping, 197–198
Carrier DNA, 140
cDNA, 23, 36, 40–44, 46–47, 84–85, 89, 139, 183, 191, 199–202, 205, 211, 217

For Product Safety Concerns and Information please contact our EU
representative GPSR@taylorandfrancis.com
Taylor & Francis Verlag GmbH, Kaufingerstraße 24, 80331 München, Germany

www.ingramcontent.com/pod-product-compliance
Ingram Content Group UK Ltd.
Pitfield, Milton Keynes, MK11 3LW, UK
UKHW021006180425
457613UK00019B/825